A SECOND VOICE

A SECOND VOICE

*A Century of Osteopathic
Medicine in Ohio*

Carol Poh Miller

OHIO UNIVERSITY PRESS   ATHENS

Ohio University Press, Athens, Ohio 45701
www.ohio.edu/oupress
© 2004 by Ohio University Press
Printed in the United States of America
All rights reserved

13 12 11 10 09 08 07 06 05   5 4 3 2 1

Frontispiece: an early osteopathic manipulative treatment, date unknown. *Courtesy of Still National Osteopathic Museum, Kirksville, Missouri* [2003.29.01K]

Library of Congress Cataloging-in-Publication Data

Miller, Carol Poh, 1950-
  A second voice : a century of osteopathic medicine in Ohio / Carol Poh Miller.
     p. cm.
  Includes bibliographical references and index.
  ISBN 0-8214-1593-X (cloth : alk. paper) — ISBN 0-8214-1594-8 (pbk. : alk. paper)
  1. Osteopathic medicine—Ohio—History. I. Title.
  RZ325.U6M54 2004
  615.5'33'09771—dc22

                                                            2004018294

The writing and publication of this history have been generously underwritten by major grants from the Brentwood Foundation and the Ohio Osteopathic Foundation, with additional support from:

Doctors Hospital (Columbus)
Doctors Hospital of Stark County
Firelands Regional Medical Center
Osteopathic Heritage Foundations

CONTENTS

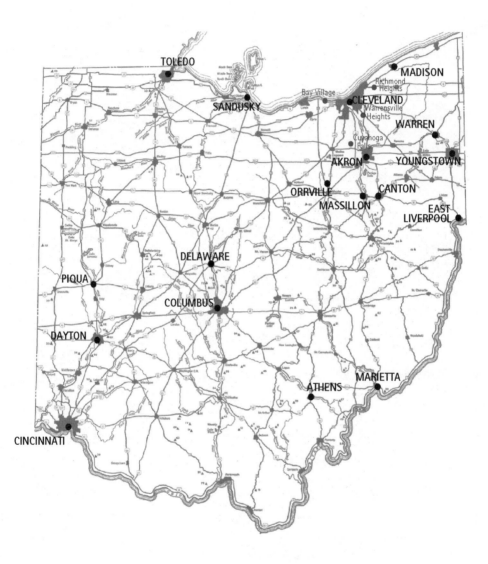

Map showing places notable in the history of osteopathic medicine in Ohio.

FOREWORD

"There is nothing more difficult to take in hand, more perilous to conduct, or more uncertain in its success, than to take the lead in the introduction of a new order of things."

So wrote Machiavelli in *The Prince* (1532). And that is what the first D.O.'s—Doctors of Osteopathy—set out to do, in the late-nineteenth century, when they bravely fought to establish a new, even revolutionary method of healing first promulgated by Dr. Andrew Taylor Still.

As the 108th president of the American Osteopathic Association and past president of the Ohio Osteopathic Association, it is indeed a pleasure and honor to introduce *A Second Voice: A Century of Osteopathic Medicine in Ohio*. Author Carol Poh Miller has masterfully documented the story of Ohio D.O.'s as they sought to "initiate a new order of things"—to establish an alternative to orthodox medical practice and, thereby, a new health care profession.

A Second Voice traces the establishment and growth of osteopathic medicine in Ohio, beginning in 1898, when thirteen osteopathic physicians met in Columbus to form a professional society, the predecessor of today's Ohio Osteopathic Association (OOA). It chronicles the profession's early hardships and its legal battles, proving that a well-organized minority is often stronger than a disorganized majority. It is a story of vision, perseverance, and personal sacrifice as early practitioners sought to gain unlimited practice rights and to establish osteopathic hospitals where they could practice without intimidation. And it is a story of professional unity, proving that an organization like the Ohio Osteopathic Association (OOA) can achieve what individuals cannot do alone.

A Second Voice documents the foresight, the determination, and the legacy of our pioneering doctors, who faced discrimination and even prosecution as they sought to establish their new order of American medicine. Here is Hugh Gravett of Piqua, the target of Ohio's early medical establishment and its advocates in state government. Here is J. O. Watson, a towering figure, who worked tirelessly to protect and expand the legal rights of the profession. Here is Richard A. Sheppard, a dedicated builder whose life and work were undone by the malicious prosecution of his son Sam for a murder he did not commit. (When you read of Richard Sheppard's good works—what he represented and what he sought to do—you will gain a wholly different perspective on the Sheppard name.) These are but a few of the profession's heroes whose stories are told in these pages—stories of men and women, doctors and laypersons dedicated to building a profession and expanding choice within American health care.

As I read this book, I was struck by a recurring theme: unity of purpose. As a struggling, minority profession, we have had to work together in order to succeed—and we have. We worked together to build hospitals and to win equal treatment under the law. In 1943, we fought for equal practice legislation and demonstrated our commitment to quality medicine by self-imposing the first mandatory continuing education licensure requirement for any profession in Ohio statutes. (Other physicians licensed by the Ohio State Medical Board did not have this requirement until 1975!) Perhaps the crowning example of our unity of purpose was the establishment of the Ohio University College of Osteopathic Medicine. By imposing an unprecedented six-year, $250 assessment on our members to support the college, the OOA demonstrated to the state legislature that Ohio's osteopathic physicians were ready to stand up and be counted.

Although the book focuses on the historical experience of Ohio D.O.'s, it illustrates the struggles common to osteopathic medicine in all fifty states. And the struggle continues today as the American Osteopathic Association seeks recognition and practice rights in foreign countries around the world.

Osteopathic medicine has been and can continue to be a meaningful "second voice" in helping to shape the American health care system. The Institute of Medicine, which is the catalyst for reforming American medicine today, has called for a new health care system for the twenty-first century focused on patient-centered care. It is our tradition of holistic medicine and primary care, together with doctor-patient communication and "hands-on" medicine, that is the essence of high-quality patient care. *Treating People, Not Just Symptoms* is more than the profession's motto—it is central to what D.O.'s do well, and it will be the foundation of high-quality health care in the twenty-first century.

In closing, I congratulate Carol Poh Miller on her excellent book. By helping us to look back, she has also helped us to look forward.

George Thomas, D.O., FACOFP
President, American Osteopathic Association (2004–5)

PREFACE

As executive director of the Ohio Osteopathic Association (OOA) since 1977 and public relations director for two years prior to that, I have witnessed many changes in the osteopathic profession. I saw the creation of the Ohio University College of Osteopathic Medicine (OU-COM) and watched as the college evolved into one of the finest primary care medical schools in the United States. I saw osteopathic unity at its finest with the establishment of the innovative Centers for Osteopathic Research and Education (CORE), and in the 1990s I was invigorated by the high-profile leadership of Barbara Ross-Lee, D.O., the first African American woman to serve as dean of a U.S. medical school. I have seen the passing of some of the profession's brightest lights—doctors like J. O. Watson, Ralph Licklider, Don Siehl, Mary Theodoras, Jerry Finer, and Jack Brill—and felt the keen sense of loss among Ohio D.O.'s as hospital sales, mergers, and closings changed the landscape of osteopathic medicine in our state. Finally, in 1998 and 1999, I was privileged to preside over the OOA as the Ohio profession marked its one-hundredth birthday.

The idea for this book was conceived in 1997 as we began planning for our centennial. A committee composed of past presidents and other key representatives of the profession convened under the able leadership of former OOA Executive Director Dick Sims. We agreed that the yearlong celebration would span two conventions, beginning in June 1998 and ending in June 1999; that it would build on the theme "A Distinguished Past— A Dynamic Future"; and that, among other special projects, we would commission a history of osteopathic medicine in Ohio that might serve as a lasting tribute to those who paved the way for today's D.O.'s.

Dick Sims was determined to get the history project moving in the right direction. For months he combed through board minutes and other documents, recording notable events that occurred prior to and during his long tenure at OOA. He brought the finished manuscript to my office, dropped it on my desk, and said, "There—now it's up to you to write your memoirs from this point forward." But memoirs were not what I had in mind (although I hope to accept Dick's challenge some day). Ohio D.O.'s have a great story to tell, and I was convinced that we needed someone from outside the profession to tell that story objectively.

Thanks to the wonders of the Internet, the search for an accomplished author to write this chronicle quite literally spanned two continents. With the assistance of OU-COM faculty member Norman Gevitz and OU-COM Director of Communications Carl J. Denbow, we posted an advertisement for a medical historian on an electronic bulletin board. Proposals soon came in from across the United States and as far away as Great Britain. We reviewed the résumés and prioritized them; I conducted interviews by e-mail and by telephone. But when the logistics of writing history from afar proved insurmountable, the search began anew to find an Ohio author.

With the help of the National Council on Public History, the trail finally led to Carol Poh Miller of Cleveland. Although not a medical historian, she was uniquely qualified to write this work. Her career as a historian and author spans more than twenty-five years. She is the author of several privately published organizational histories, as well as coauthor of *Cleveland: A Concise History, 1796–1996* (Bloomington: Indiana University Press, 1997). She also had an intense curiosity about the osteopathic profession.

From the outset, Miller's charge was to write a scholarly, objective history that would be thoroughly documented, factually accurate, and readable—a work written to appeal not only to D.O.'s, their employees, and their patients, but also to the educated general reader. "An important objective of the project, as I see it," she wrote in her letter of proposal, "would be to make a solid contribution to a heretofore-neglected aspect of the history of medicine in Ohio." In *A Second Voice: A Century of Osteopathic Medicine in Ohio* she has done just that. Borrowing from our centennial theme, it is my sincere belief that the distinguished past documented in these pages will help light the way toward a dynamic future for osteopathic medicine in Ohio.

Jon F. Wills

ACKNOWLEDGMENTS

Writing this history required substantial self-education, and the task would have been impossible without the kind assistance of Jon Wills, executive director of the Ohio Osteopathic Association, to whom I owe special thanks. In personal meetings and by telephone, letter, and e-mail, Jon helpfully and patiently answered my questions about the osteopathic profession when he could and quickly referred me to others when he could not. He put the OOA archives at my disposal, helped track down elusive source material and photographs, and offered the informed perspective that can come only from someone who has devoted nearly three decades to the osteopathic profession in Ohio. Jon's entire staff, especially Cheryl Markino, was also unfailingly kind and helpful whenever I visited the Columbus headquarters office.

I also wish to thank the members of the manuscript advisory committee, who each read multiple drafts and made valuable comments. In addition to Jon Wills, they are Norman Gevitz, Carl Denbow, Dick Sims, and Mary Jane Carroll. Ms. Carroll's father, William S. Konold, was a central figure in the story of osteopathy in Ohio, and she generously shared personal files related to his life and work. I am also grateful to Philip K. Wilson, historian of medicine and associate professor of humanities, Penn State College of Medicine, who read an early draft of this work and offered many valuable comments.

Many others assisted me in small but meaningful ways. Rick Vincent, president of the Osteopathic Heritage Foundations, read the manuscript and shared his perspective on the rash of hospital closings and consolidations that occurred in the 1990s, and its meaning for the profession. Chip Rogers, director of advocacy for the Ohio University College of Osteopathic Medicine, gave me an extended, informative, and enjoyable tour of the Athens campus. Fran Collins, head of the Memorial-Nottingham Branch of the Cleveland Public Library, arranged convenient access to the library's comprehensive holdings of the *Buckeye Osteopathic Physician* and its predecessor, the *Buckeye Osteopath*, key resources for my work. William C. Barrow, special collections librarian at Cleveland State University Library, and Debra Loguda-Summers, curator of the Still National Osteopathic Museum in Kirksville, Missouri, helped locate and copy several photographs in the book. Marilyn M. Cryder, curator of the Delaware County Historical Society, provided historical information about the Delaware Springs Sanitarium, the first osteopathic hospital established in Ohio, and its successor, the Delaware Osteopathic Hospital. Ida Sorci, coordinator of the library and archives of the American Osteopathic Association, tracked down biographical information for numerous early Ohio doctors. Finally, two people helped me revisit the painful case of Dr. Sam Sheppard. Sam Reese Sheppard, Dr. Sheppard's only child, shared recollections of his January 2000 address to the Cleveland Academy of Osteopathic Medicine, and Dr. Stephen A. Sheppard, Sam Sheppard's brother, granted permission to quote liberally from *My Brother's Keeper,* his 1968 book about the murder of Marilyn Sheppard and its aftermath.

Lastly, I wish to thank Rick Huard and Beth Pratt of the Ohio University Press for their care and creativity in bringing this work to print.

INTRODUCTION

A NEW SCHOOL OF MEDICINE

> The human body is a machine run by the unseen force called life,
> and that it may be run harmoniously it is necessary that there be
> liberty of blood, nerves, and arteries from their
> generating point to their destination.
>
> —*Andrew Taylor Still*

Osteopathy was born of tragedy and heartache, the brainchild of a man who lost three children, in swift succession, to spinal meningitis. Andrew Taylor Still (1828–1917) would build on a truth posited by Hippocrates—that illness is not the work of the gods but the result of a malfunction of the body—by focusing on the unity of all body parts and identifying the musculoskeletal system as a key element of health.

Born in Jonesville, Virginia, Still was the son of an itinerant preacher and doctor who ministered to the Indians and pioneers of the western frontier in Missouri and Kansas.[1] His medical education was limited to self-study, experience gained working at his father's side, and service as a hospital steward and surgeon during the Civil War. The medical horrors of wartime caused Still to become dissatisfied with orthodox medical practice. Then, in February 1864, he lost three children to spinal meningitis. His heart "torn and lacerated with grief," Still blamed the gross ignorance of the medical profession and determined to find a better method of healing.

Medical practice in nineteenth-century America was primitive, with poorly trained practitioners employing harsh therapies in an attempt to cure diseases they barely understood. Bloodletting and toxic pharmaceuticals were commonly used, while useful agents, such as digitalis, quinine, and opium and its derivatives were commonly misused; the germ theory—the connection between particular infectious ills and particular microorganisms—had yet to be firmly established. One historian has written, "Armed with cups, lancet, and leech and provided with calomel, tartar emetic, arsenic,

Andrew Taylor Still, 1895. *Courtesy of Still National Osteopathic Museum, Kirksville, Missouri [PH 692].*

and an assortment of other drugs, doctors proceeded to bleed, blister, puke, purge, and salivate patients until they either died from the combined disease and treatment or persevered long enough to recover from both."[2] Not surprisingly, the inadequacies and hazards of traditional medicine gave rise to a host of alternative methods of healing, including eclectic medicine, homeopathy, hydropathy, mesmerism, phrenology, and Christian Science. Among the ranks of nineteenth-century reformers who sought another way, Andrew Taylor Still developed his own unique medical philosophy.

Working in a home laboratory in Baldwin, Kansas, Still immersed himself in research, conducting extensive anatomical dissections of humans and animals. On June 22, 1874, he later wrote, "I was shot—not in the heart, but in the dome of reason. . . . I began at that date to carefully investigate with the microscope of mind to prove an assertion that is often made . . . that the perfection of Deity can be proven by His works. . . . I began to look at man. What did I find? I found myself in the presence of an engine—the greatest engine that mind could conceive of."[3] Believing that he had found the key to health and disease in the perfection of the body—with which the use of drugs was fundamentally incompatible—Still thereafter divorced himself from traditional medicine.

Still's abandonment of drugs and his exploration of alternative theories of healing led to his ostracism in Baldwin. He moved his family to Missouri, eventually settling in Kirksville, seat of Adair County. There, he refined his philosophy, basing it on evolutionary principles and close study of the nerves, spinal cord, and brain. Still's revolutionary theory: that most diseases were directly or indirectly caused by vertebral displacements whose elimination through spinal manipulation would allow the unobstructed flow of blood, remove symptoms of pathology, and enhance the body's natural curative powers. The "Lightning Bone Setter"—so Still advertised himself—traveled across Missouri, preaching his message to ever-larger crowds. Dispensing with the use of drugs, he treated headache, heart disease, paralysis, sciatica, rheumatism, and a host of other ailments by manipulating vertebrae to "put them into the proper place." In 1885, he finally named his new science of healing, combining *osteon* (bone) with *pathos* (suffering) to create osteopathy.

As Still's reputation grew, patients descended on the little town of Kirksville, arriving on foot and horseback, by passenger train and in covered wagons, filling Still's office to capacity. When three patients asked to study with him, Still agreed. Confident that his science and technique could be taught, in the fall of 1892, at age sixty-four, Still opened the American School of Osteopathy (ASO) in a two-room wood-frame building. Its purpose was "to improve our present systems of surgery, obstetrics, and treatment of disease generally, to place the same on a more rational and scientific basis, to impart information to the medical profession and to grant and confer such honors and degrees as are usually granted and conferred by reputable medical colleges."[4] Upon completion of several months' course work in anatomy, osteopathic principles, and technique, students received the degree of Diplomate (later Doctor) of Osteopathy (D.O.). The first class of five women and sixteen men—including three of Still's own children and one of his brothers—was graduated in 1894.

To answer his critics and obtain equal treatment under the law, Still lengthened the course of study to two years in 1896 and expanded the curriculum to include every subject covered in a standard medical college with the exception of materia medica;[5] osteopathy would be a drugless healing art. As rival schools sprang up elsewhere, passage of Missouri's osteopathy law in 1897, giving D.O.'s the right to treat disease through the hands, brought hundreds of new students to Kirksville. Graduates of the "Missouri

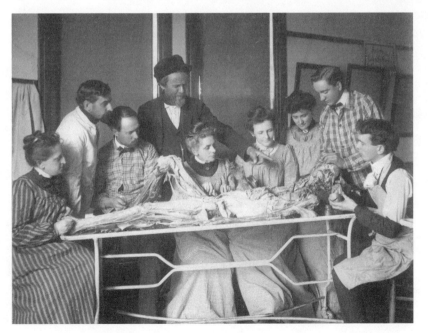

Andrew Taylor Still conducts an anatomy class, about 1897. *Courtesy of Still National Osteopathic Museum, Kirksville, Missouri [PIC-DIS-1].*

His favorite study: Andrew Taylor Still examines a femur, about 1910. Although his health had begun to fail, Still continued to write on a variety of subjects until he was debilitated by a stroke in 1914. *Courtesy of Still National Osteopathic Museum, Kirksville, Missouri [PH 641].*

Mecca" went forth to establish their own private practices. By 1900, forty-eight osteopathic physicians had settled in Ohio.

The "Old Doctor," as Still was fondly known, spent his later years trying to preserve the purity of his osteopathic doctrine. He continued to reject drug therapy. But the scientific achievements of Koch, Pasteur, and Lister could not be ignored, and by the time of Still's death, in 1917, an enlarged ASO faculty had integrated bacteriology and immunology into the now four-year curriculum, thereby broadening the intellectual base Still had laid. Although osteopathy would come to employ the diagnostic and curative tools of scientific medicine, Andrew Taylor Still's contributions would endure. He had not only focused attention on the role of the musculoskeletal system and the value of manipulative therapy in patient care; he had given rise to a new health-care profession having a focus on the whole person.

OHIO'S OSTEOPATHIC PIONEERS

Osteopathy, in its essence, is radically different from all
former practice of the healing art.
—*E. R. Booth*, History of Osteopathy, *1924*

OSTEOPATHY IN OHIO TRACES ITS origin to 1896, when Eugene H. Eastman, D.O., opened an office in Akron. About the same time, Herman Taylor Still, D.O., the son of Andrew Taylor Still, opened offices in Hamilton
and Cincinnati, and Grace Huston of Circleville enrolled in the American
School of Osteopathy in Kirksville, becoming Ohio's first osteopathic
medical student.[1] Other D.O.'s who established offices in Ohio before 1900
included: in Cleveland, Frank G. Cluett, Therese Cluett, George J. Eckert,
Helen Marshall Giddings, and Charles McLeod Turner Hulett; in Cincinnati, George W. Sommer and Charles A. Ross; in Columbus, Mac F. and
Adelaide S. Hulett; in Piqua, Hugh H. Gravett; in Springfield, Mary A.
Connor, L. H. McCartney, and J. T. L. Morris; in Toledo, William J.
Liffring; and in Warren, James F. Reid. Among this group, the number of
women is notable: denied entrance to most medical schools, they were wel-
comed into the osteopathic colleges.[2]

Many of Ohio's early D.O.'s were attracted to osteopathy because of
satisfactory results that they, a friend, or a family member had received at the
hands of an osteopathic physician. Charles A. Ross, D.O. (1872–1958), an

HULETT C. M. TURNER, D. O.,
AND NELL MARSHALL GIDDINGS, ,
D. O., Osteopathists, 1208 New England
bldg.; Tel. Main 3102; Hours: Monday,
Tuesday, Thursday, Friday, 9 to 4,
Wednesday and Saturday, 9 to 12
— Harry C. (Johnson & Hulett), 329 Central av.
— Ralph M. architect, 451 Pearl.r.165 Marvin

The Cleveland city directory
of 1900 included this listing
for two of Ohio's early osteopathic physicians, Charles
McLeod Turner Hulett,
D.O., and Nell (Helen) Marshall Giddings, D.O. *Courtesy
of Cleveland Public Library.*

1899 graduate of the American School of Osteopathy who practiced in Cincinnati, recalled that his sister Laura had been given a "death warrant" from the family doctor treating her for tuberculosis, but was subsequently successfully treated by osteopathic physicians. Effie B. Koontz, D.O. (1868–1956), a native of Missouri, became interested in osteopathy through a friend who had been healed by osteopathic treatment. Koontz sought treatment for a twisted spine from Dr. Andrew Taylor Still at Kirksville. For six months, she was a patient in his home. With her health restored, Koontz enrolled as a student, studying under Dr. Still in the small two-room building that housed the first school. In 1901, Koontz established an office in London, Ohio, where she practiced for more than fifty years.[3]

Another early physician, William J. Keyes, D.O. (1872–1945), who practiced in Portsmouth and, later, Norwood, Ohio, suffered a neck injury in 1892 that left him in chronic pain. In 1895, he went to Kirksville for treatment and found relief. His experience, together with the testimony of others he knew who had also benefited from osteopathic treatment, inspired him to enroll in ASO, from which he was graduated in 1900. An injury likewise led Katherine McLeod Scott, D.O. (1873–1959), to osteopathy. The native of Brunswick, Canada, was on the deck of a sailboat when it collided with another vessel, whose jib-boom swept her into the sea. After spending three years in a wheelchair, in desperation she sought treatment from a doctor who had studied with Dr. Still at Kirksville; within six months she was walking. After receiving her D.O. degree from ASO in 1905, Scott established an office on High Street in Columbus with her husband, John Herbert Brice Scott, D.O., where they practiced for fifty years.[4]

The first D.O.'s treated disease using only their hands. They believed that the healing powers of manipulation went far beyond the correction of structural problems or dislocations—that the perfect adjustment of all parts of the body cleansed the blood, rejuvenated the glandular system, and helped the body to resist disease. Indeed, manipulation alone was probably as good as, or better than, many orthodox medical therapies available at the time; patients suffering from chronic ailments often sought out the D.O. after finding no relief from traditional medicine. But the task of educating the public about osteopathy could be daunting. In 1900, Therese Cluett, D.O., of Cleveland, reported her experiences in dealing with ignorance and misunderstanding of her medical art. "A lady entered my office and asked if I was a theosophist. I said, 'No madam, I am an osteopathist.' 'Oh well,' she replied, 'It's the same thing.'" Another patient asked Cluett if she was a Christian. "This fairly caught my breath. . . . I asked her if she had put the same question to [her last physician] that she had put to me. She replied that she had not." Cluett reported that she spent an hour explaining osteopathy to each of these patients.[5]

As they struggled to establish their practices, the first D.O.'s often suffered great personal hardship, and the conditions they faced in the field could be primitive. Especially in rural areas, early osteopathic physicians conducted itinerant practices, traveling widely, without benefit of autos or highways, to visit their patients. Hugh H. Gravett, D.O. (1862–1957), who came to Ohio in 1897, established offices in Piqua and Greenville. In addition to his office practice, he responded to calls as far away as Cincinnati and Toledo, traveling at night and on Sundays and holidays. He would later recall that "there was at that time possibly not half a dozen people in the entire county who knew anything of Osteopathy. If they did, they were afraid to say so." His first prospective patients warned him that they "would have to be a little guarded about letting people, especially the doctors, see us go to your office or having you come to our homes." Then, the minister of the

First D.O.'s Faced "Trouble Aplenty" but Set "Remarkable Record"

William E. Reese, D.O., of Toledo, was honorary chairman of the 1951 convention of the Ohio Osteopathic Association. In his remarks at the president's banquet, Reese (1872–1968), who began his practice in 1905, recalled his early experiences in the profession.

It might seem that my practice was established the hard way. It did not seem hard to me. It was my desire to eat regularly, so an office was maintained in Bowling Green three alternating days in the week, to pay the rent of the Toledo office occupied the other three days of the week.

Commuting was by interurban. Calls were made on foot, bicycle, horse and buggy and streetcar. Days spent in Bowling Green began at 6:30 A.M. and usually terminated with the arrival of the last interurban—around midnight.

I consider myself neither a patriarch nor a pioneer in the profession. But we had trouble aplenty back in 1905. X-ray facilities were not available to us. Hospitals were a closed door. Consultation with specialists was denied. Laboratory work was nil. Nurses were hostile. A referred surgical case was equivalent to—goodbye patient.

It was not unusual to be requested to enter the home of an influential patient by means of the rear door, along with the plumber and meter reader. The over-cautious patient requested secrecy lest neighbors and friends discover he was experimenting with unorthodox treatment . . .

During the First World War, when influenza, accompanied by pneumonia, was rampant, the osteopathic profession across the country established an enviable record. In Toledo, out of hundreds and hundreds of cases treated by our profession, but one was lost. Nurses were unobtainable. In many instances our physicians served in the capacity of doctor, nurse, and cook. On numerous occasions in critical cases I have remained throughout the night lest some members of the family stampede and—in their terminology—"call a doctor."

You may well ask what medication was used to achieve this remarkable record. Penicillin was not known. Oxygen tents were not in existence. The medication, ladies and gentlemen, consisted of the *ten thinking, seeing, feeling fingers* of the osteopathic physician.

church dissuaded Gravett and his family from attending services, explaining that "there seems to be considerable opposition to it on account of your being an Osteopath."[6] An early husband-and-wife team in Cleveland, William H. Schulz, D.O., and Pearle Baker Schulz, D.O., recalled that, with no access to hospitals, they sometimes resorted to using the kitchen table, the dining room table, even the ironing board to administer treatment and perform minor surgery. Coupled with such hardships was the fear of prosecution. "The word 'responsibility' had a very grave meaning then, before we had the Major Surgery licenses," according to the Schulzes. "Always hanging over one's head was the thought of standing before the Judge and answering his question of 'Did you or did you not?'"[7] In fact, no sooner did the first D.O.'s hang out their shingles than they attracted the attention of local prosecutors.

Medical practice in Ohio was essentially unregulated until 1896,[8] when the Medical Practice Act was passed, establishing a state board of medical registration and examination. It authorized the governor to appoint seven members, with schools of practice to be represented "as nearly as possible in proportion to their numerical strength in the state, but no one school to have a majority of the whole board."[9] The first board consisted of three allopathic (orthodox) physicians, two homeopathic physicians, one eclectic physician, and one physiomedical physician; within a year its makeup had changed to three allopathic, two homeopathic, and two eclectic practitioners.[10] Nothing in the law governed osteopathy. Not surprisingly, the new school with its revolutionary approach provoked a severe reaction among the medical establishment and its political representatives. Ohio D.O.'s—D.O.'s in every state—would fight a protracted battle for legislative recognition and equal practice rights.

The first osteopathic physician in Ohio was also the first to be prosecuted.[11] In 1897, Dr. Eugene Eastman was found guilty in Akron mayor's court of practicing medicine without a license. The 1896 Medical Practice Act stipulated: "Any person shall be regarded as practicing medicine or surgery within the meaning of this act who shall append the letters M.D. or M.B. to his name, or for a fee prescribe, direct or recommend for the use of any person, any drug or medicine or other agency for the treatment, cure or relief of any wound, fracture or bodily injury, infirmity or disease."[12] Eastman appealed the decision in Summit County Common Pleas Court. Although Judge J. A. Kohler ruled in his favor, finding that his "particular acts . . . do not constitute a violation of this statute," Eastman soon left the state.

On the heels of this ruling, the Ohio Attorney General joined the campaign against the state's osteopathic physicians. In November 1897, Assistant Attorney General John L. Lott advised U.S. Senator Joseph B. Foraker, Republican of Ohio and a staunch defender of osteopathy, of his intentions. "We are very anxious to have the question [of osteopathy's legal status] settled," he wrote, "and I think I shall within the next two or three weeks, cause an arrest to be made, and have the question determined as speedily as possible."[13] William J. Liffring, D.O. (1869?–1945), a graduate of the Northern Institute of Osteopathy (Minneapolis) practicing in Toledo, was indicted by a grand jury and arrested for practicing medicine without a license. The defendant entered a demurrer, which was sustained by the Lucas County Common Pleas Court.[14] The case was then carried to the Ohio Supreme Court, where the lower court's findings were sustained in November 1899.

With prosecution threatening every osteopathic physician in the state, Ohio D.O.'s recognized the need to work collectively for mutual protection and professional advancement. On the morning of December 31, 1898, thirteen of the fifteen osteopathic physicians then practicing in Ohio met in the Columbus office of Drs. Mac and Adelaide Hulett and organized the Ohio Association for the Advancement of Osteopathy. Membership would be open to graduates of "any reputable school of osteopathy" residing in Ohio. The first officers were George W. Sommer, D.O., Cincinnati, president; P. F. Kirkpatrick, D.O., Columbus, vice president; Mac F. Hulett, D.O., Columbus, secretary; and William J. Liffring, D.O., Toledo, treasurer. Reporting on the meeting, the *Columbus Dispatch* noted that Dr. Hulett "entertained the doctors at the Neil [Hotel] at noon and this evening Dr. Kirkpatrick will give them a complimentary dinner at the Chittenden."[15]

In March 1900, the Ohio House of Representatives amended the state's Medical Practice Act by passing the "Love Act." Rep. Maro J. Love, Republican of Erie County and chairman of the House Medical Committee, authored the legislation. Ostensibly written to *protect* the rights of osteopathic physicians, the law as amended in fact contained a provision with which no osteopath could comply. "This act," it read, "shall not apply . . . to any osteopath who holds a diploma from a legally chartered and regularly conducted school of osteopathy in good standing as such, wherein the course of instruction requires at least four terms of (5) five months each in four separate years."[16] Mac Hulett would later write, "There being no such school of osteopathy, the osteopaths refused to accept this subterfuge, and continued to practice as before."[17]

Mac F. Hulett, D.O., date unknown. On December 31, 1898, thirteen of Ohio's fifteen osteopathic physicians met in Hulett's Columbus office to organize the Ohio Association for the Advancement of Osteopathy, predecessor of the Ohio Osteopathic Association. *Courtesy of Ohio Osteopathic Association.*

The Love Act, which provided no representation for osteopathic physicians on the state medical board and imposed educational requirements no osteopathic physician could meet, was a clarion call to the fledgling profession. It would need to act to ensure its own self-preservation. On May 19, 1900, members of the Ohio Association for the Advancement of Osteopathy met at the home of Mac Hulett and unanimously agreed to ignore the law and resist all attempts to enforce it. A few days later, Hulett wrote to every D.O. in the state, announcing the establishment of a judiciary committee and pledging financial assistance to any member in good standing.

No osteopath applied for a license under the Love Act, which the state medical board proceeded to enforce. Following the death of a Greenville woman, the *Piqua Daily Leader* on October 16, 1900, reported the arrest of "H. H. Gravett, the Osteopath" for practicing "in violation of the law enacted by the last legislature." Gravett, an ASO graduate and a charter member of the Ohio Association for the Advancement of Osteopathy, was subsequently indicted by a Darke County grand jury.[18]

Gravett had come late to the case of the Greenville woman, Mrs. E. D. Huddle, after two medical doctors had tried, unsuccessfully, to treat her for an unspecified illness. Following newspaper reports of Gravett's indictment, the woman's husband quickly came to the doctor's defense, asserting his belief that, had Dr. Gravett received the case sooner, his wife would yet be living. "It was not the people of Greenville that caused Dr. Gravett's arrest," the Piqua paper quoted Huddle as saying, "but it is the work of the medical fraternity for their own selfish interest." Gravett's attorney, A. F. Broomhall, entered a demurrer, which was sustained by the Darke County Common Pleas Court. By consent of both parties, the case was carried to the Ohio Supreme Court.

Two issues were at stake: whether osteopathy was the practice of medicine under the law; and whether the osteopathic amendment contained in the Love Act, allegedly made to *regulate* osteopathy, in fact *prohibited* it and therefore was unconstitutional. On December 3, 1901, the Ohio Supreme Court sustained the demurrer. "Osteopathy wins," Broomhall telegraphed his client. Hugh Gravett, who would practice into his ninetieth year, was long

```
Ohio Association for the Advancement
           of Osteopathy.
                      Secretary's Office.

                      Columbus, Ohio, May 23, 1900.

Dear Doctor:-
          A called meeting of the Ohio Association was held at
Columbus , May 19th, to discuss the attitude of Osteopathy toward
the New medical law.     Following is a report of the business
transacted.
          It was unanimously agreed to completely ignore the State
Medical Board of Registration and Examination as having any claims
upon us, and to resist to the highest courts if necexxary any at-
tempt to enforce the recently enacted law to the exclusion of os-
teopathy.     Should you be notified to appear for examination, pay
no attention to it.
        · The Association will lend its financial assistance only to
members in good standing and regular graduate osteopaths who con-
tribute their share of the expense, upon becoming members.
          The Secretary was authorized to draw on each member by sight
draft at any time it may be needed, in sums the aggregate of which
shall not exceed $25.00 per member, to create a fund to defend
against prosecution under the medical law.
          In all future meetings of the Association. R.R. Fares of those
in attendance shall be divided equally.
          The report of the legislative contest as read by the Secretary
was ordered revised (if needed) by a committee, and printed for
general distribution among our friends, each member to receive 25
copies, and as many more as he may desire at his own expense. (Sam-
ple proof sheets will be furnished as soon as possible.     Upon
receipt of same please notify the Secretary how many extras you
desire.)
          Dr. N.O. Minear having left the state, his office as member
of the Executive Committee was declared vacant, and Dr. E.W. Goetz
was appointed to fill the vacancy.
          A Judiciary Committee, was appointed, consisting of Dr. E.W.
Goetz, Neave Bldg., Cincinnati; Dr. C.M.T. Hulett, New England Bldg.
Cleveland, and the Secretary of the Association.     In case of
prosecution or threatened prosecution, or if you know of any other
information of similar interest, communicate with this Committee
at once.
          All osteopaths of the State not members how were urged to
become such as soon as possible.
          A vote of thanks was extended to Mrs. M.A. Shoup, of Carmi,
Ill.,-but a native of Ohio- for her efficient assistance in present-
ing the subject of Osteopathy to the last General Assembly.     Mrs.
Shoup came to Columbus on her own solicitation especially for this
purpose, staying through the entire session at her own expense.

                      Very truly,

                      M.T. Hulett, D.O., Sec'y.
```

Letter from Mac F. Hulett, D.O., secretary of the Ohio Association for the Advancement of Osteopathy, announcing the establishment of a fund to defend members in good standing against prosecution under the odious Love Act. *Courtesy of Ohio Osteopathic Association.*

hailed by the profession as a pioneer in the battle to obtain legal recognition for osteopathy in Ohio. On the inside cover of a scrapbook he kept of news articles pertaining to his arrest and prosecution, Gravett would later scrawl, "As I now read these clippings they appear quite amusing but at the time this little drama was being enacted I assure you it gave no such pleasure."[19]

Within days of the Gravett victory, Ohio's osteopathic physicians met at the Chittenden Hotel in Columbus for their fourth annual convention. The mood was jubilant as Hugh Gravett, then serving as president, addressed the gathering. The organization adopted a new constitution changing its name to the Ohio Osteopathic Society. The articles of incorporation stipulated the organization's purpose:

OSTEOPATHY WINS.

Case Against Dr. H, H. Gravett]Decided in His Favor By the Supreme Court To-day.

The Supreme Court today handed down a decision that was an important one to every physician practicing Osteopathy, in that it gives them the privilege of following their profession without violating the state law.

It was the decision in the case of Dr. H. H. Gravett of this city, who was arrested last fall on complaint of a citizen of Greenville on the charge of practicing medicine without a license. It was carried to the Supreme Court, and this afternoon Dr. Gravett received the following message from his attorney.

Osteopathy wins. Decided this morning.   A. F. BROOMHALL.

The case has been attracting a great deal of attention all over the state..

Article in the *Piqua Daily Call* from the scrapbook kept by Hugh H. Gravett, D.O. Gravett was the first osteopathic physician to be prosecuted under the Love Act. *Courtesy of Ohio Osteopathic Association.*

To seek to promote the interests & influence of the science of Osteopathy & of the Osteopathic profession by all means that will conduce to their development & establishment such as:

The stimulating & encouraging of original research & investigation & the collecting & publishing of the results of such work for the benefit of the whole profession.

The elevation of the standard of Osteopathic education & the cultivating and advancing of Osteopathic knowledge & the fostering & directing of a correct public opinion as to the relations of practitioners of Osteopathy to Society & the State & providing for the united expression frequently and clearly of the views of the profession thereon.

The promoting of friendly emulation & social intercourse among the members of the profession & of prompt & intelligent concert of action by them in all matters of common interest.[20]

The Gravett decision left the practice of osteopathy in Ohio unregulated, and in the next session of the Ohio General Assembly, the state's osteopathic physicians proposed their own amendment to the medical law. Arthur G. Hildreth, D.O. (1863–1941), a member of the faculty of the American School of Osteopathy who had helped secure passage of the Missouri law, came to Ohio to lobby for the bill. After much hard work by both Hildreth and Mac Hulett—to whom Hildreth assigned "more credit for our success . . . than to any other man"—it passed unanimously in the house and encountered only four dissenting votes in the senate. Gov. George K. Nash signed the bill into law on April 21, 1902.[21]

The legislation amended the Medical Practice Act to create a three-member Osteopathic Examining Committee to examine applicants in the subjects of pathology, physiological chemistry, gynecology, minor surgery, osteopathic diagnosis, and principles and practice of osteopathy. If successful, the applicant was then examined by the state medical board. Persons already engaged in the practice of osteopathy in Ohio, and who held a diploma from "a regular college of Osteopathy as determined by the committee," could apply for a certificate from the medical board without examination. The certificate would allow the holder "to practice osteopathy in the state of Ohio, but shall not permit him to administer drugs nor to perform major surgery."[22]

The new law was hailed as a victory for the profession; Hildreth judged it "one of the best laws if not the best now in existence."[23] But while the amended law gave legislative recognition to osteopathy in Ohio, it did not recognize osteopathy as a complete school of medicine, and the limitations it imposed would stymie the profession for the next forty years.

2

THE STRUGGLE FOR RECOGNITION AND EQUAL PRACTICE RIGHTS

It is utterly impossible to conceive that our school can live very long
unless our practitioners are given unlimited right to heal the sick.
—*F. C. Smith*, D.O., Journal of Osteopathy, *1916*

Walter H. Siehl, D.O., of
Cincinnati, chronicled the
profession's early history
and was the patriarch of an
extended osteopathic family.
This photograph was taken
shortly after his election as
president of the Ohio Osteopathic Association in 1948.
*Courtesy of Ohio Osteopathic
Association.*

WITH THE PASSAGE OF THE LAW giving recognition to osteopathy in
Ohio, members of the profession turned their focus to "public relations and
to [building] up their private practices," according to one early Cincinnati
practitioner, Walter H. Siehl, D.O. (1888–1955). "It was now necessary to
establish their bases, cement their gains and improve their technology,"
Siehl wrote, and to "impress upon their patients and friends the value of
their services." Siehl, however, also recognized that osteopathy's victory
was less than complete. For even as the new law gave confidence and the assurance of protection to the osteopathic profession, it revealed "the need
for additions, corrections and legal interpretations . . . in order to perma-
nently establish Osteopathic-medicolegal prerogatives."[1]

As the profession sought to win public acceptance and expanded practice rights, it faced formidable opposition from orthodox medicine. Under the guise of enforcing the law and
protecting the public from impostors, the American Medical
Association and its constituent societies lobbied to prevent the
recognition and licensure of "irregular" healing groups, including osteopathic physicians. A resolution adopted in 1902
by the Miami County Medical Society, an affiliate of the Ohio
State Medical Association, illustrates the prevailing climate:

WHEREAS, There is now pending in the Ohio Legislature a bill known as the "Brown Bill," to establish a board of
examiners to legalize a pretended system of curing disease

by rib adjustment, spine setting, bone pulling, nerve pressing, and pipe adjusting, claimed to have been discovered by H. E. Still [*sic*], of Baldwin, Kansas, in 1894, known as Osteopathy,

WHEREAS, This class of men and women are manifestly ignorant of the first principles of a medical education and totally ignorant of the nature of disease. . . . Be it therefore

Resolved, That it is the sense of this Society that to legalize this class of pretenders, thereby opening the doors of the sick room to them, would be a serious menace to the health of the community. . . . Be it further

Resolved, That it is the consensus of opinion of this Society that the system has no foundation based on experience or good sense, and therefore [is] to be classed with the Indian Hoo-doo or Doweyite, or Christian Science pretenders, wholly unscientific and therefore dangerous to the commonwealth.

In *History of Osteopathy,* Emmons R. Booth, D.O., of Cincinnati, reprinted the resolution to illustrate the "extreme measures" to which the "medical fraternity" often resorted. He then dismissed the document, writing, "Such statements . . . would excite mirth if they did not arouse the higher feeling of pity that a noble profession should be so willing to prostitute the truth." Indeed, he added, the resolution was so reactionary that it "did Osteopathy good rather than harm."[2] Despite harassment from "regular" medicine, the osteopathic profession in Ohio gradually gained ground. In 1902, Florence L. McCoy, D.O., of Toledo, became the first osteopathic physician to give expert testimony in a court of law. In 1905, a ruling by the state attorney general gave osteopathic physicians the right to sign death certificates. And in 1910, a ruling by the state attorney general upheld the right of osteopathic physicians to treat contagious or infectious diseases.[3]

Reconstructing the early history of osteopathic medicine in Ohio is unfortunately hindered by the loss of the profession's earliest records. We know that, by 1911, small numbers of D.O.'s had established practices in the largest cities. An Ohio Osteopathic Society membership directory published that year presents a picture of the fledgling profession. There were then eight D.O.'s in Cincinnati, twelve in Cleveland, five in Columbus, and nine in Toledo. There were four D.O.'s in Canton and three each in Akron, Dayton, and Marion. The remaining members were scattered about the small towns of Ohio, singly or in pairs: Ada, Bowling Green, Bucyrus, Chillicothe, Delaware, East Liverpool, Elyria, Findlay, Fremont, Galion, Greenville, Hamilton, Hicksville—102 in all.[4]

Profile
Emmons R. Booth, D.O.

In January 1934, the *Buckeye Osteopath* reported the death of "one of our best loved pioneer physicians." Emmons Rutledge Booth (1851–1934), an educator-turned-physician, made a lasting mark on osteopathic medicine both in Ohio and nationally. A native of Indiana, Booth taught school in St. Louis and served as president of the Missouri State Teachers Association before enrolling in the American School of Osteopathy at Kirksville. An obituary would later explain that he was inspired to become an osteopathic physician when his wife, "a sufferer for years . . . was evidently saved from the grave by osteopathy." Booth received his D.O. degree in 1900 and, at age forty-nine, established a medical practice in Cincinnati.

In 1901, Booth was named the first chairman of the Osteopathic Examining Committee of the

Emmons R. Booth, D.O., of Cincinnati, was a distinguished and beloved member of the early profession in Ohio. His classic *History of Osteopathy* provides a window into osteopathy's early struggles. *Courtesy of Ohio Osteopathic Association.*

Ohio Board of Medical Registration and Examination. He served a term as president of the American Osteopathic Association (1901–2), and in 1903 the AOA and the Associated Colleges of Osteopathy appointed him as the first college inspector. Booth's assignment was to make an on-site survey of each of the nation's osteopathic colleges. His findings and recommendations persuaded the AOA to require that all colleges establish a compulsory three-year, twenty-seven-month curriculum. While the requirement caused some colleges to fold because of declining enrollment—one was the short-lived Ohio College of Osteopathy in Chillicothe, in operation from 1902 until 1905[1]—the addition of a third year helped improve the standing of others. By 1915, seven recognized D.O.-granting schools remained, located in Boston, Chicago, Des Moines, Kansas City, Kirksville, Los Angeles, and Philadelphia. (As directed by the AOA, all of the colleges added a compulsory fourth year in 1916.)

In 1905, Booth made a singular contribution to the profession, publishing his exhaustive *History of Osteopathy, and Twentieth-Century Medical Practice,* a volume he revised and enlarged in 1924. Dedicated to Andrew Taylor Still, a personal friend, the book chronicled the development of osteopathy, its schools, its legal battles, and its principles and practice. Booth minced no words. A historian, he explained, "has the right to state facts with which he is familiar, and may be pardoned if he uses language intended to make his meaning unmistakable."

On the establishment of state and local osteopathic societies:

> The early opposition of the drug doctors to the osteopaths in most states made it necessary for the latter to organize in order to present a united front against a united and a thoroughly organized and trained foe. The osteopaths in some of the states were slow to appreciate the advantages of state societies and a close alliance with the national association. They soon found, however, that a single-handed fight was generally a losing one.

From the beginning, Ohio was well represented in the profession nationally, furnishing two presidents of the American Osteopathic Association (AOA)—Charles McLeod Turner Hulett in 1900–1 and Emmons R. Booth in 1901–2—and hosting three AOA conventions prior to 1920: at Cleveland in 1903, Put-in-Bay in 1906, and Columbus in 1917.[5] At the Put-in-Bay meeting, delegates arrived by steamboat from railheads in Detroit, Cleveland, Toledo, and Sandusky. The convention banner draping the

The Ohio Osteopathic Society Directory of 1911 listed 102 members, one quarter of whom were women. *Courtesy of Ohio Osteopathic Association.*

ℭᵭ BUCKEYE OSTEOPATH

*" * * * with an eye single to the glory of Osteopathy"*

Vol. I   DAYTON, OHIO, AUGUST, 1923   No. 2

PRESIDENT'S MESSAGE
R. A. Sheppard, D. O.

Inasmuch as I was not present at the election of the new President of the Ohio Society of Osteopathic Physicians and Surgeons, I take this my first opportunity to formally thank you for the honor thus bestowed upon me. I keenly feel the responsibility that accompanies this important position. I cannot hope to measure up to the ability and efficiency of our past president, who has done so much to organize the state into a working unit. However, I will endeavor, with the help of my able officers and the co-operation of the society as a whole, to carry on the work which has been started by those who have preceded me. I regret very much not being able to be present the last day of our annual state meeting. I am sure you will excuse my absence when you realize it was necessary for me to leave so that I might arrive in Boston on time to enter the Post Graduate class conducted by Robt. H. Nichols. It was a most profitable month's work and I recommend the taking of this course by all, either at Boston or Delaware.

I had the pleasure of spending three days at the National Convention at New York. I wish it might have been possible for all of the Ohio Osteopaths to have been there. It was one huge success in every detail and was conceded by all with whom I spoke to be the best convention osteopathy has ever put on. I was delighted to see the large representation from Ohio and I am proud that Ohio, "the home of Presidents," has within her midst a brand new president, in no other than our efficient state secretary, Dr. W. A. Gravett, who carried off the Presidency of the A. O. A. We are very glad that one of our members should be so honored, and I am sure we are all ready to co-operate in any way to help him during his administration.

In regard to our Ohio state work for the coming year. The present administration, in mapping out their program or policy, has felt it better to concentrate their efforts on one or two things of utmost importance, and put them across, rather than spread their energy over a

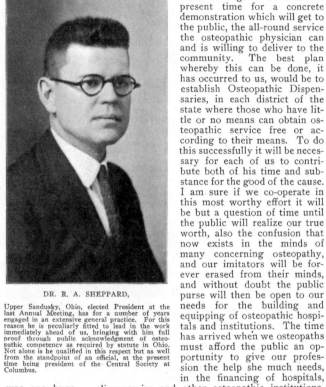

DR. R. A. SHEPPARD,

Upper Sandusky, Ohio, elected President at the last Annual Meeting, has for a number of years engaged in an extensive general practice. For this reason he is peculiarly fitted to lead in the work immediately ahead of us, bringing with him full proof through public acknowledgment of osteopathic competency as required by statute in Ohio. Not alone is he qualified in this respect but as well from the standpoint of an official, at the present time being president of the Central Society at Columbus.

number of endeavors. We feel there is a great need at the present time for a concrete demonstration which will get to the public, the all-round service the osteopathic physician can and is willing to deliver to the community. The best plan whereby this can be done, it has occurred to us, would be to establish Osteopathic Dispensaries, in each district of the state where those who have little or no means can obtain osteopathic service free or according to their means. To do this successfully it will be necessary for each of us to contribute both of his time and substance for the good of the cause. I am sure if we co-operate in this most worthy effort it will be but a question of time until the public will realize our true worth, also the confusion that now exists in the minds of many concerning osteopathy, and our imitators will be forever erased from their minds, and without doubt the public purse will then be open to our needs for the building and equipping of osteopathic hospitals and institutions. The time has arrived when we osteopaths must afford the public an opportunity to give our profession the help she much needs, in the financing of hospitals, dispensaries and other osteopathic institutions. So far as I know, we have no institutions organized wholly to serve the community without profit. We must give the people an opportunity to help in this work. It is only human nature to be interested and strive for the success of these things in which we have invested our time and money. The medical fraternity understands this trait of human nature and works for it to the Nth degree. So we must awaken in the public an interest in our undertaking. We must solicit and obtain their support for osteopathic charitable institutions, such as the dispensaries which we are about to establish in each district of Ohio. Before we approach the public for support each of these dispensaries must be organized and functioning. This work will unite us as a profession in a definite community service. It will be the

(Continued on page two)

Launched in 1923, the *Buckeye Osteopath* sought to unify the struggling minority profession. "This little paper has a mission," the *Buckeye* of October 1934 declared. "Those who support it have the satisfaction of knowing that their science of osteopathy is being defended." *Courtesy of Ohio Osteopathic Association.*

Hotel Victory—said to be the largest summer hotel in America—bore the inspiriting words of Oliver Hazard Perry: "We have met the enemy and they are ours." It was at the Put-in-Bay meeting that the AOA took the first step toward establishing an institution for "scientific investigation and research" of osteopathic technique. C. M. Turner Hulett, D.O., of Cleveland, led the fund-raising campaign, and the A. T. Still Research Institute finally opened in Chicago in 1913.[6] At the Columbus convention, in August 1917, scientific sessions shared the agenda with politics: a bill was then pending in Congress to make licensed osteopathic physicians eligible for medical commissions in the armed forces, and D.O.'s hoped to help supply "the enormous demand for specialists for the reconstruction of crippled soldiers at home and in Europe."[7] From the program for that Columbus meeting in 1917, we know that there were then 213 osteopathic physicians in Ohio, of whom 149 belonged to the state society.[8]

William A. Gravett, D.O., the first editor of the *Buckeye Osteopath*, served as president of the Ohio Osteopathic Society in 1915–16 and as president of the American Osteopathic Association in 1923. *Photo by Cornwell–Dayton, courtesy of Ohio Osteopathic Association.*

As the ranks of osteopathic physicians grew larger, the Ohio Osteopathic Society was able to undertake new initiatives. Beginning in 1916, news of interest to the profession was reported in the pages of the *Bulletin of the Ohio Osteopathic Society,* edited by Frank A. Dilatush, D.O., and published in Cincinnati. In 1923, the publication was relaunched as the *Buckeye Osteopath,* edited by William A. Gravett, D.O., and published quarterly in Dayton. The few surviving issues present a picture of a small but determined minority profession eager to safeguard the welfare of its members and to promote osteopathy as a progressive school of medicine.

In 1916, a legislative committee consisting of Drs. E. R. Booth (Cincinnati), F. C. Smith (Marion), and M. F. Hulett (Columbus) called on society members to pledge a small amount of money each month. The proceeds enabled the Ohio Osteopathic Society to retain Columbus attorney Clarence L. Corkwell—"a clean cut, energetic and likable young man"—as legal counsel. (Corkwell's brother, F. E. Corkwell, D.O., practiced in Newark.) "Now," declared the *Bulletin of the Ohio Osteopathic Society,* "every osteopath in the state is represented in legal affairs by one man, who is making it his particular business to acquaint himself with the osteopathic point of view and to know every phase of the laws under which we practice."[9]

Corkwell immediately went to work, and in 1917, following "a very difficult battle with our medical opponents," the society secured a change in state law permitting osteopathic physicians to use antiseptics and anesthetics, and to perform minor and orthopedic surgery. Two years later, a bill to

In Their Own Words
Early D.O.'s Tell How It Was

In 1973, Loren D. Leidheiser, D.O., of Huron, asked his senior colleagues to share their reminiscences about the early practice of osteopathic medicine in Ohio. Leidheiser (1911–1989), a 1935 graduate of Kirksville College of Osteopathic Medicine and a longtime OOA trustee, hoped to publish a history of the profession. That hope was not realized, but the replies he received from all corners of the state provide a vivid picture of osteopathic medicine as practiced in the first half of the twentieth century.

It was rough going in 1930; my office calls were $2.00. . . . Having no hospital connections, I delivered babies in the homes, in the country, in trailers, even in a chicken coop where a family was living. . . . Before the advent of wonder drugs, I treated pneumonia with manipulation and encouraged good nursing care. Today I do the same but also combine the wonder drugs. In either case I have never lost a pneumonia patient and I am proud of this record.

Marie A. Keener, D.O., Canton

O.B.'s were delivered in the home, sometimes by the light of an oil lamp, but [there was] never a serious postpartum infection. I quit practicing obstetrics when going to the hospital became popular and I was refused hospital privilege[s].

I remember treating a lady in her seventies, for bronchopneumonia without sulfas, antibiotics or even oxygen. . . . By treating her every eight hours, relaxing the muscles in the thoracic and cervical areas, raising the ribs and even using the lymphatic pump whenever possible, we pulled her through and she lived for ten more years.

R. E. Hofmeister, D.O., Hicksville

I opened my office for practice in Feb. 1925. At that time a majority of patients came to me because of the failure of orthodox medicine. . . . I handled pneumonia, scarlet fever and all acute diseases. The worst case of Rheumatic fever I have ever seen was handled osteopathically. The man is living today and there was no residual heart disease. . . . Fees in those days were $2.00 for office calls and $2.50 for house

calls. . . . Obstetrical cases were handled in the home. . . . Surgical cases were taken to the Bashline-Rossman Hospital in Grove City, Pa. Forty-two miles away.

Harry E. Elston, D.O., Niles

Before sulfas and antibiotics, I treated pneumonia by manipulation, with great success. In fact, it is my opinion that this is the choice treatment for pneumonia today if we could get our hospital staffs to do [it].

John W. Mulford, D.O., Cincinnati

When I started in practice March 5, 1935, I had no patient the first day, twenty-three the first month. My fee was $1.50 for office calls and $2.00 for house calls. I collected $937.00 the first year plus $30,000 worth of experience. . . . We treated our patients with the ten fingers before antibiotics were discovered and I think I can verify we had a low percentage in the mortality column.

Lobar Pneumonia patients, we would ventilate the lungs by placing the patient on his or her back and gently raising the rib cage and let it drop, thus causing the air to rush into the lungs. We would continue this treatment until a cold sweat was noted on the face and then stop for a period of 15 minutes and repeat.

Warner S. Eversull, D.O., Cincinnati

When a freshman at Ohio Wesleyan it was my good fortune to get a job in the bath department at the Delaware Springs Sanitarium which enabled me to stay in school. Equally important, my contact with Dr. L. A. Bumstead, and others, was my first step in the road to Kirksville. Four years later by the grace of God . . . I received a diploma which proclaimed me to be a physician. Only I knew how colossal was the exaggeration.

Returning to Ohio in January 1929 I was employed by the Marietta Osteopathic Clinic at $125.00 per month. It was a living of sorts but more important my five years there provided experience and much needed confidence. . . . My first night call made me feel quite important. The glamour faded however when I had to follow a lantern through a cornfield to the Ohio River and walk a plank to a house boat. No pay. . . . In 1935 I joined Dr. John W. Hayes in East Liverpool. One year later we began with six beds the development of our present small but modern hospital. . . . Only the D.O.'s who practiced a generation or more ago can appreciate the changes. We are no longer referred to in our local paper as "West Fifth Street Osteopath." We are no longer compared with M.D.'s as "osteopaths and doctors."

Opposition has contributed to our strength.

C. M. Mayberry, D.O., East Liverpool

How could I forget those first three months! On October 1, 1927, I came to Columbus and moved into a 6-room apartment over a drug store and grocery store. We lived in 4 rooms and I had my office in the other 2 rooms. The first 3 months I took in $78. Being a graduate Pharmacist I decided to look for part time work. . . . Fortunately, in 4 months I began to do enough so I could pay expenses and did not have to get a job.

Before the miracle drugs, sulfas and antibiotics, we treated many cases of influenza and pneumonia. Yes, we made house calls and when someone called a D.O. they expected an Osteopathic treatment. We all developed bedside technique. The one I can remember most was in 1928. I was called at 11 P.M. to see an 11-month-old baby after an M.D. had just left and given it up to die. The child was in a cold sweat, the temperature was subnormal. It was hoarse when it cried and its lungs were filled with rales. I stayed all night. Every hour . . . I raised its ribs and gave it lymphatic pump and gentle OMT. The child lived and it was a victory for Osteopathy. . . . In 1932 I was given O.B. privileges in the old Radium Hospital. One could put an O.B. patient in for 10 days and the charge for everything in the hospital was [a] $40 flat fee. . . . In 1940 Dr. J. O. Watson, Dr. Harold E. Clybourne and Dr. Ralph S. Licklider purchased Doctors Hospital which had 32 beds. Anyone who wished to be on the staff pledged $500 to help remodel it. That was the start of the wonderful hospital we have now. It really gives me a thrill to look down the long hallways now as I make my rounds every day. I am forever grateful to the three Doctors and to William S. Konold who helped make it possible.

Ralph T. Van Ness, D.O., Columbus

I graduated from KCOS in 1938. Following graduation I interned at Marietta, Ohio, for one year. I was married in 1939 and spent one year in Zanesville as an assistant to a well-established D.O. I worked on a salary basis and received $25.00 a week. . . . After leaving Zanesville, I went into private practice in Newcomerstown, population 4,500. . . . There were 7 M.D.'s in the town and I was definitely not accepted. I charged $1.50 for an office call and $2.50 for a house call. The first month I took in $60.00. . . . My first O.B. was a home delivery in the country. This was with the help of one oil lamp. When the husband would leave the room for some purpose the lamp would be taken with him and we were in complete darkness. Fortunately, it was an uncomplicated delivery and she knew more about having babies than I did. My fee at that time was $25.00 for pre-natal care and the delivery. Most deliveries were in the home with the husband giving a little ether. . . . I was not accepted in the county medical hospital but did have privileges in a privately owned medical hospital at Cambridge, Ohio. . . . About 18 years ago a county medical hospital was built in Cambridge. I was a charter staff member and have been an active staff member ever since with the same privileges as any of the M.D.'s. . . . Before we had the sulfas and antibiotics I treated pneumonia and other infections with osteopathic manipulation treatment. Sometimes treating them two or three times a day. I was fortunate and did not lose any cases of pneumonia. We had more to offer than the M.D.'s at that time and of course still do.

R. L. McCullery, D.O., Newcomerstown

I originally had made arrangements to enter Case Western Reserve Medical School but prior to starting, I received a basketball injury to my left hand which became infected and I suffered from a general septicemia. . . . [An M.D.] was called in as our attending physician. After about six or eight weeks in bed with no improvement (there being no antibiotics of any kind at that time), my mother insisted on calling in a local D.O. who had recently started a practice in Barberton, Ohio. This man was Dr. Samuel Lash. . . . Despite the fact that I resisted a "rub doctor" treating someone like me who was very ill and sore all over, a decided change in my condition was noted after only three treatments.

Needless to say, I soon became interested in a therapy that had worked when allopathic medicine had failed and later became interested in Kirksville College of Osteopathy and Surgery from which Dr. Lash had graduated. . . . I returned to Akron to practice where I have remained ever since.

I started practicing on July 29, 1929, and the stock market failed as well as the banks in August. I paid $25.00 a month rent for my office and had very difficult times meeting this obligation. . . . There was no such thing as insurance coverage for office calls, hospitals, or anything at that time and many days I was elated when I collected $2.00 from a patient. . . . Practically all of our work, both in home calls and the office was manipulation. However, we did have many venereal disease cases, which helped "tide" us over the rough spots when they would pay. . . . I think I began to get a good general practice when in desperation a family in Kenmore, a part of Akron, called me to see a baby only a few days old with bronchopneumonia. The attending medic was unable to obtain any results with medication and I spent many days and nights treating the patient with osteopathic therapy, especially the "rib-raising technic," and the baby made good progress and recovered. Needless to say those patients and their offspring have been osteopathic patients ever since.

Most insurance companies would not accept examinations with a D.O. signature. . . . I did secure good rapport with the coroner's office, the health department, etc., but the school board would not accept a student's excuse for illness with the signature of an osteopathic physician until later. . . . I had an active obstetrical practice in those days in the home. . . . There were no hospital privileges at that time. . . . We would take our surgical cases, when we could get them to go up to Cleveland where Dr. Richard Sheppard practiced and, before him, Dr. A. C. Johnson. Also, there was an osteopathic surgeon who came through the state, not on horseback, but in an old car and we would try and have surgery scheduled for him in nursing homes, etc. around town.

In 1943 I was able, with three other D.O.'s in the area, to obtain an old hospital in East Akron which was run by a nurse. We waited until she went into bankruptcy and then bought her equipment and rented the building. The hospital started with thirty-five beds and then later fifteen more were added in the basement and due to the atmosphere which was that of a "big family," we prospered all those years until we were able to start the present Green Cross General Hospital in Cuyahoga Falls.

Arthur L. Harbarger, D.O., Cuyahoga Falls

Almost 45 years ago when I began my practice in East Liverpool, Ohio, things were far different than they are now. . . . Generally, I was not accepted by the public at large but, being indoctrinated and well so in the practice of osteopathy, which in those days was limited to the 10 fingered variety, I plugged along and in due time was better accepted. . . . In those early days I was, of course, criticized and objected to by many in the allopathic profession. One prominent physician in this city was attempting to have my license removed because I was attending a case of scarlet fever. At least two of my friends in the medical profession . . . attempted to correct this individual and, apparently, they did for I still have my license. . . . Some of us well remember that when we were licensed, our license said "osteopathy and surgery" and it meant just that. We were permitted to use manipulative therapy, to do surgery, to perform deliveries, but we could not prescribe any medication of any kind except antiseptics and anesthetics. How did we get along in those days? Strange as it may seem, many of our cases of flu, pneumonia, or what have you, recovered and miraculously so, with just manipulative therapy and good common horse sense. . . . I well remember the time that unfavorable legislation was about to be introduced against the osteopathic profession in Columbus. A call went out from our State Secretary, who at that time [1934] was M. A. Prudden, D.O. of Fostoria, and on a Sunday morning more than 600 of the practicing members of the osteopathic profession converged on Columbus. . . . At least one of the legislators was known to have said that they could not do this to this profession because we were there ready to fight. . . . In those days we were fighting for our very livelihood and now, the young men in practice now, apparently, have no need to worry for those older practitioners of the profession in Ohio have certainly paved the way and paved it very well.

John W. Hayes, D.O., East Liverpool

add major surgery to the subjects in which osteopathic physicians were to be examined by the state medical board carried unanimously; the bill also included an amendment to Section 1288 of the General Code changing the word "Osteopath" to the more dignified "Osteopathic Physician."[10]

During its early years, the society's annual meetings rotated to cities around the state, relying on local physicians to plan and host the gatherings. By 1923, when the society again changed its name—to the Ohio Society of Osteopathic Physicians and Surgeons—six district organizations were firmly in place, in Columbus, Dayton, Cincinnati, Toledo, Akron, and Cleveland. Most were large and active enough to hold one program each month. The Cleveland District Osteopathic Society, for example, in 1924 sponsored symposia on diabetes and tuberculosis, with selected doctors presenting case reports. In addition to such local programs, a statewide Lyceum Bureau, chaired by John J. Coan, D.O., of Cleveland, arranged for osteopathic specialists in and outside of Ohio to visit each district organization to lecture on advanced medical topics—an early forerunner of continuing medical education.

When the *Buckeye Osteopath*, after a brief hiatus, made its second debut, in September 1933, just over four hundred osteopathic physicians were practicing in Ohio. Supported by advertising and edited by Arthur Collom Johnson, D.O., a respected Cleveland orthopedic surgeon, the monthly publication was mailed to every osteopathic physician in Ohio, as well as to other state osteopathic organizations, 138 daily and weekly newspapers, and Ohio lawmakers. Billing itself as the "Watchdog of the Profession," the *Buckeye* presented the image of an embattled profession that daily faced the enmity of orthodox medicine.

In his first issue, Johnson skewered the AMA for its oft-repeated claim that "the Osteopaths are trying to get into medicine by the back door," labeling it a "bullism."[11] Noting the demise of homeopathy ("We recently had occasion to search through the telephone directory for a list of Homeopathic Physicians. Not a name could we find!"), Johnson cautioned osteopathic physicians against being likewise absorbed into "the mess of Allopathic pottage."[12] And, asserting professional pride, Johnson declared, "We know of 405 Osteopathic Physicians in the State, and we don't know of a single one who is a quack. What a tremendous argument that is as we seek the good will of the public! What other profession can say as much?"[13] The publication's tone apparently startled some readers, leading its editor to confess that "our advertising man thinks we are too pugnacious in our attitude toward those who don't like us" and to promise to "add a little more weight to the arm of our safety-valve. The pressure sometimes gets so high it simply blows off in spite of all we can do to hold it in."[14] Johnson nevertheless remained unfettered. In a 1934 editorial articulating the challenges to osteopathic medicine, he declared, "Whether osteopathy can retain its identity throughout the years to come, is the vital question. Whether osteopathy has made itself, as a separate and distinct method for the treatment of disease, secure against the noxious designs of the antagonistic dominant school of medicine, remains to be seen." The time is ripe, he concluded, "for us to bare our fangs and defend our place in the sun."[15]

By 1920, the hospital had become an accepted part of medical and especially surgical care for rural and urban Americans alike.[16] Denied privileges at most allopathic hospitals by the rules of the American Hospital Association, osteopathic physicians were forced to establish their own institutions. Typically, these began in converted residences or office buildings, with more space added as demand grew. They represented the vision and the sacrifice of dedicated individuals, usually surgeons, who needed a place to operate and to serve their patients. Most were private; few had the benefit of either public funds or private philanthropy.

The first osteopathic hospital in Ohio, the Delaware Springs Sanitarium, was founded in 1914 as a stock company by L. A. Bumstead, D.O., a graduate of the American School of Osteopathy. As construction proceeded on a new facility on fifteen rural acres of high, rolling land in the Olentangy River Valley, the sanitarium opened in a private residence in downtown Delaware, a sedate college town and trading center. J. H. Long, D.O., a Kirksville classmate of Bumstead's who had gone on to study at Harvard, joined the staff as general surgeon, and here in this makeshift hos-

Opened in 1916, the Delaware Springs Sanitarium was the first osteopathic hospital in Ohio. This postcard view dates from the 1920s. *Courtesy of Delaware County Historical Society.*

pital Long performed the first operation by an osteopathic surgeon in Ohio.[17] In March 1916, the new sanitarium, built at a reported cost of $92,000, was formally opened with a public reception. The three-story English Tudor-style building boasted thirty private rooms as well as a general ward; in the basement, a hydrotherapy room was designed to take advantage of seven mineral springs located on the grounds.[18]

The osteopathic profession in Ohio quickly responded to support the new facility. A waiting list for sanitarium and surgical care prompted the construction of a fifty-bed addition in 1920. The pioneering institution became a focus for osteopathic medical education, hosting an "Osteopathic Chautauqua" organized by Reginald H. Singleton, D.O., of Cleveland, and postgraduate courses in "better diagnosis" sponsored by the Ohio Society of Osteopathic Physicians and Surgeons. Although a slumping postwar economy and overwhelming debt forced the institution into receivership in 1926, Bumstead later recalled with pride the many contributions of the Delaware Springs Sanitarium. "More than eighty percent of Ohio osteopathic physicians referred patients to the sanitarium," he wrote, "which was an outstanding record. Training of nurses and interns was a valuable service to our profession. Many of our prominent osteopathic practitioners in Ohio served their internship in the Delaware Springs Sanitarium." The early surgical work done there, Bumstead noted, "had a great influence in securing the law permitting osteopathic physicians to include major surgery in their practice."[19]

In 1921, Drs. Heber M. Dill and Frank A. Dilatush established the Dill-Dilatush Clinic in Lebanon, between Dayton and Cincinnati. Emmons Booth described it as "sanitary, well lighted, pleasantly furnished, and fully equipped with everything necessary for the small sanitarium or hospital. Many minor operations are performed. The service is osteopathic first, last, and all the time."[20] Four years later, Dill and Dilatush, together with William A. Gravett, D.O., established the ten-bed Dayton Osteopathic Hospital in a former private residence at 325 West Second Street, near the downtown business district. In 1935, the hospital was reorganized under Drs. Dill, Dilatush, and Richard F. Dobeleit, a surgeon and radiologist. Incorporated as a nonprofit organization in 1940, the hospital would grow rapidly following World War II and its relocation to a new site on Grand Avenue, taking the new name of Grandview, in 1947.[21]

In southeast Ohio, the Marietta Osteopathic Hospital was an outgrowth of the Marietta Osteopathic Clinic, opened in 1927 by Drs. Hubert L. Benedict, J. E. Wiemers, L. M. Bell, and J. D. Sheets. When Marietta Memorial

The Dayton Osteopathic Hospital, opened by William A. Gravett, D.O., in 1926, typified the beginnings of many osteopathic hospitals, which got their start in converted residences. *Photo by Horstmann Studio, Dayton, courtesy of Ohio Osteopathic Association.*

Hospital refused to permit osteopathic surgeons to work in its facility, the same four doctors organized a hospital on December 30, 1934, forming a nonprofit corporation with two hundred stockholders. A two-story, twenty-five-bed fireproof building, connected to the rear of the old wood-frame clinic, was formally opened in 1935. Five years later, it was enlarged following a substantial gift from Francis M. Selby, a wealthy independent oil producer whom Dr. Benedict had successfully treated for a knee ailment. The new addition increased hospital capacity to fifty beds and included a large solarium equipped and furnished by the women's auxiliary.[22]

Operating room, Dayton Osteopathic Hospital, about 1940. *Courtesy of Ohio Osteopathic Association.*

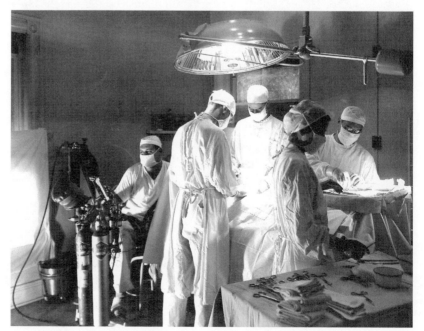

The Marietta Osteopathic Clinic, opened in 1927, became a hospital in 1935 with the addition of a twenty-five-bed fireproof building at the rear of the clinic. *Courtesy of Ohio Osteopathic Association.*

In 1938, the *Buckeye* announced the addition of Memorial Hospital in Sandusky to the state's osteopathic institutions. Several years earlier, D.O.'s Lester R. Mylander, C. W. Koehler, and O. C. Ricelli had converted the first floor of the Loretta Britton Convalescent Home, a cobblestone residence built in 1910 for the Hinde family, for the care of obstetrical patients. As other area D.O.'s began to use the facility, the convalescent home relocated to new quarters and the small osteopathic hospital expanded to two floors, with fourteen beds, five bassinets, and operating and delivery rooms. In 1940, the hospital was incorporated as a nonprofit institution under the governance of a board of trustees.[23]

A solarium was part of a major expansion of Marietta Osteopathic Hospital made in 1940 thanks to a generous gift from businessman Francis M. Selby, whose photograph graces one of the tables. "Vita Glass" allowed ultraviolet light to penetrate the windows. *Photo by Walter C. Tracy, courtesy of Ohio Osteopathic Association.*

In Cleveland, meanwhile, a group of doctors led by surgeon Richard A. Sheppard, D.O., embarked on one of the period's most ambitious projects when they purchased and renovated a mansion at 3146 Euclid Avenue. Built in 1870 by piano dealer George Hall as part of the city's once-famed Millionaires' Row, the three-story mansion represented Sheppard's vision to create "one of the outstanding clinics for his profession in the country." The Cleveland Osteopathic Hospital opened in September 1935. By the close of 1939, the hospital had admitted

Memorial Hospital in Sandusky (*at right*) and the Cleveland Osteopathic Hospital both opened in former private residences in the mid-1930s. *Courtesy of Ohio Osteopathic Association.*

3,067 patients and its surgical staff had performed 1,784 operations. In September 1941, Sheppard added a twenty-room nurses' residence and a public dispensary for the care of indigent patients.[24]

The Hayes-Mayberry Osteopathic Hospital in East Liverpool received a charter to operate as a nonprofit corporation in 1940. The incorporators were surgeon John W. Hayes, D.O., and eye-ear-nose-throat specialist

Doctors Hospital, Columbus, 1940. The Second Empire–style house, which once housed Radium Hospital, was the nucleus of what would become one of the Ohio profession's most successful institutions. *Courtesy of Ohio Osteopathic Association.*

C. M. Mayberry, D.O., together with their wives. The hospital's beginnings were modest: three beds, presently increased to six, on the second floor of an office building at 142 West Fourth Street. In 1946, the corporation reorganized and moved the hospital to larger quarters in a Queen Anne-style house on West Fifth Street. There, bed capacity gradually increased and the facility changed its name to the East Liverpool Osteopathic Hospital.[25]

Perhaps no osteopathic hospital in Ohio enjoyed greater success than Doctors Hospital, founded in the spring of 1940 by a group of Columbus osteopathic physicians and surgeons headed by Ralph S. Licklider, D.O., James O. Watson, D.O., and Harold E. Clybourne, D.O. The group purchased the former Columbus Radium Hospital, a three-story Victorian house on Dennison Avenue, and promptly embarked on a $40,000 remodeling program. "All Central Ohio is excited with pride over the prospect of an actual osteopathic hospital in the Capitol City," the *Buckeye* reported in July 1940. When the renovated Doctors Hospital opened its doors to patients in August 1940, its thirty-four beds and ten bassinets filled quickly. Despite the exigencies of wartime, the hospital embarked on the construction of a two-story north wing in 1944–45, bringing capacity to sixty-four beds and twelve bassinets. Yet another expansion project, financed by subscription, added a new south wing and more than doubled capacity, to 150 beds, in 1948.[26]

In addition to these pioneering institutions, D.O.'s during this period could also be found at approximately twelve public hospitals in those Ohio communities where local statutes prohibited discrimination.[27] But even where osteopathic physicians enjoyed access to hospitals, the prejudice they faced could be formidable. One D.O. reported that, while attending an obstetrical patient at the Salvation Army's Booth Memorial Hospital in the Cleveland suburb of Euclid, he was informed that Booth was a "Class A hospital" according to the American College of Surgeons, and therefore no osteopathic physician or surgeon was "fit" to practice there.[28] John W. Keckler, D.O., who practiced at Cleveland's Roscoe Osteopathic Clinic in the 1920s and 1930s, recalled that, although clinic doctors had access to a few of the city's smaller hospitals, including the East 79th Street and Willson Street Hospitals, cooperation from the medical profession was so poor that "it became our aim . . . to handle all the cases we possibly could at the office or [patient's] home."[29]

In 1939, the Ohio Osteopathic Hospital Association (OOHA) was organized to promote the influence and science of osteopathy and the osteopathic profession "by establishing, extending, and maintaining high standards of hospital service."[30] Dr. Richard Sheppard was elected as the first president of the group. It was a small but meaningful measure of the profession's maturity. "Can you see the importance of the osteopathic hospital to you?" Paul L. Riemann, superintendent of Marietta Osteopathic Hospital, asked *Buckeye* readers. "Can you see the importance of the osteopathic hospital to the profession? The hospital is lifting the osteopath out of the back rubbing category, into a class on a par with medical doctors." He urged D.O.'s to send their patients to the nearest osteopathic hospital "even if it is 200 or 300 miles away."[31]

As the profession grappled with prejudice and worked to establish hospitals, it also looked for new ways to win friends for osteopathy. In 1933, Dr. Richard Sheppard organized a free clinic committee, asking a trustee from each of the society's districts to serve on it with him. Sheppard proposed to hold free clinics in each of Ohio's eighty-eight counties. "I believe if we do this work successfully," he wrote in the *Buckeye*, "it will so educate the people concerning Osteopathy and its scope, that [favorable] legislation will be gained more easily. Needless to say, it will advertise the local doctor in a dignified and legitimate way."[32]

The first free clinic was held in Shelby in February 1934. Another, larger one followed in Canton at the Hotel Northern on April 11, 1934, when "scores of patients" were examined by twenty volunteer physicians. An editorial in the *Buckeye* hailed the experiment, contending that the clinics "show the public that osteopathy is not a narrow, myopic, bigoted system of therapeutics. [The clinics] tend to disperse the obscuring vapors of mystery . . . and they put osteopathy outside the pale of cultism" even as they "renew the enthusiasm of the osteopathic physicians."[33] A year later, Mark A. Bauer, D.O., secretary-treasurer of the Stark County Osteopathic Society, reported that the Canton clinic had achieved the hoped-for result. "Most of these patients have since found their [way] into the offices of the county's osteopathic physicians," he wrote in the *Buckeye*.[34]

Cleveland surgeon Richard A. Sheppard, D.O., an activist in the profession, championed free clinics as "a dignified and legitimate way" to educate the public about osteopathy. *Photo by Frank Moore Studio, courtesy of Ohio Osteopathic Association.*

In 1939, the Ohio Society of Osteopathic Physicians and Surgeons sponsored the first of what would become an ongoing series of free child health clinics at the Cuyahoga County Fair in Berea, examining 126 children, caring for 27 emergencies, and distributing 1,000 pieces of literature. The success of the effort led Reginald H. Singleton, D.O., of Cleveland, chairman of the Committee on Fairs and Expositions, to suggest that the state's D.O.'s hold child health clinics at other county fairs and at the state fair in Columbus. "Health advice given to eager parents and educational literature presented to thousands of Ohio's citizens will mean less resistance to Osteopathy in our legislative battles," he wrote in the *Buckeye*. "People could be made to realize the scope and efficiency of Osteopathy in a more practical and certain manner, than by the employment of any other measures we now have at our command."[35] Such outreach efforts, it was hoped, would educate the public about osteopathy and help disperse the "obscuring vapors" that, all too often, led confused patients to ask—to the consternation of the D.O.—"Can you give a chiropractic treatment?"[36]

With its vast social and economic dislocation, the Great Depression brought sweeping changes to the medical profession, including increased federal regulation of medical and surgical care, and the prospect that government-subsidized health care would be extended to the poor. The osteopathic profession nationwide found itself concerned with an avalanche of legislative measures affecting its rights and privileges. In March 1938, the Ohio Society of Osteopathic Physicians and Surgeons established a Committee on Maternal and Child Health, deemed "one of the most significant committees in the history of the Ohio profession." Chaired by Dr. James Watson, its purpose was to ensure osteopathic participation in new federal health

From his office at the Cleveland Osteopathic Hospital, Raymond P. Keesecker, D.O. (1891–1960), edited the profession's monthly, rechristened the *Buckeye Osteopathic Physician*, beginning in 1936. That year, only about half of the state's osteopathic physicians were members of the state professional society, which struggled financially. *Photo by Fabian Bachrach, courtesy of Ohio Osteopathic Association.*

programs by securing recognition by the Ohio Department of Public Health, through which the money would be allocated. Watson was concerned that, if osteopathic physicians and institutions were excluded, "it will simply mean that the school of practice will have a large source of its income entirely shut off." His concern was well founded. Alluding to the Burke-Drew Bill, then pending in Congress, Watson recalled that, in 1916, when a commission was set up to render service to government employees injured or taken ill while on duty, only M.D.'s were ruled eligible to render such service. "The total number to be served was small, and no one [in the osteopathic profession] bothered to even consider a protest. Thirty years later economic conditions forced the government to become the employer of millions, but a precedent established in 1916 prevented osteopathic physicians from gaining recognition as physicians for the W.P.A.!"[37]

With the relocation of its editor, Dr. Arthur Collom Johnson, to Detroit, the *Buckeye Osteopath* suspended publication following the July 1935 issue. In November 1936, publication resumed under a new name, the *Buckeye Osteopathic Physician*, with Raymond P. Keesecker, D.O., at the helm as editor. Keesecker (1891–1960) had helped found the Cleveland Osteopathic Hospital, where he headed the departments of radiology and anesthesiology.[38] "D.O.—Dig On" was the clarion call of his inaugural issue:

> Organized Osteopathy in Ohio has never faced a brighter future. Its ranks are augmented by new men, progressive, enthusiastic, militant for their rights, but osteopathic in their concept of practice. They contribute the fresh blood essential to the maintenance of a vital group. At the same time the profession still has within its ranks many of its pioneer leaders. It is essential that the vitality of youth be merged with the wisdom of experience if the profession in Ohio is to go forward, presenting a solid front to all who would undermine it.[39]

Despite the *Buckeye*'s professed optimism, osteopathic medicine in Ohio faced large hurdles. The Depression had eaten into doctors' incomes, affecting their support of and participation in professional activities. In

Gertrud Helmecke Reimer, D.O., of Cincinnati, served as president of the Ohio Society of Osteopathic Physicians and Surgeons in 1935–36. When this photograph was taken, in 1938, she was an active member of the society's Committee on Maternal and Child Health, formed to ensure osteopathic participation in new federal health programs. *Courtesy of Ohio Osteopathic Association.*

1933, just ninety-four Ohio D.O.'s attended the annual meeting in Toledo, "which number was considered very good," noted the minutes of the board of trustees, "when the financial condition of the state was considered."[40] In view of the economic hardship many were experiencing, and in an attempt to boost its rolls, the society waived its requirement that those seeking reinstatement pay a year's back dues.[41] Nevertheless, the society's membership and budget remained anemic. In 1936, only slightly more than half of the state's 441 osteopathic physicians were dues-paying members of the Ohio Society of Osteopathic Physicians and Surgeons.[42]

With the help of attorney Clarence Corkwell, the society's legislative committee continued its efforts to protect member physicians from discrimination and unfair attacks. But its efforts to obtain representation on the state medical board and equal rights under the law had been uneven and not always strongly supported by the society's own members. "There is still evidence that some of the profession are uneducated as to the needs for a better law governing the practice of Osteopathy and Surgery in Ohio," the board of trustees concluded in its year-end summary for 1928–29. "It is lamentable that . . . some physicians cannot see the position of danger that is now existing."[43]

In 1934, legislative chairman Mac Hulett reported that the committee had adopted a policy of "watchful waiting." During the past year, he wrote, "there has been no particular need for active effort for either change in legislation or defense to protect against abridgement of our present rights. . . . In our opinion, we are 'sitting pretty,' being able to practice as we see fit, the 'persecutions' of the licensed osteopathic physicians by the State Medical Board having completely fallen down." Hulett concluded that "so long as these conditions remain, and so long as the financial stress stands as it does, we are better off to undertake no radical change."[44] The society's own

Meryl A. Prudden, D.O., of Fostoria, served as secretary of the Ohio Osteopathic Society in the 1930s, a difficult period when the organization was hobbled by a lack of money and members. *Courtesy of Ohio Osteopathic Association.*

financial woes likely contributed to Hulett's caution. "Our big weakness," the society's secretary, Meryl A. Prudden, D.O., confessed, "is the lack of money to carry on the fight in a large way."[45]

Late in 1936, Ohio Attorney General John W. Bricker stunned the profession by issuing a ruling restricting the right of osteopathic physicians to administer anesthetics and antiseptics. Membership Chairman O. R. Glass, D.O., declared that the move "has set us back 25 years." At the regular meeting of the board of trustees, on September 12, 1937, Dr. James Watson, who had succeeded Mac Hulett as legislative chairman, advised a "complete reorganization" of the society's method of handling legislative affairs. A full-time executive, he said, was needed to carry on the work. Two months later, Watson repeated his plea, citing both the Bricker ruling and the proliferation of New Deal legislation, especially Social Security, now affecting the profession. "Various commissions having to do with medical relief for indigents, crippled children, the blind, etc., are multiplying rapidly," he wrote, "and, so far, there has been no one to adequately represent and protect our interests."[46]

At its midwinter meeting, the board considered Watson's recommendations that membership dues be raised to $15 and that a full-time executive secretary be hired, but failed to act on it.[47] Watson continued to hammer home the message that, in the new social order, osteopathy must be vigilant in protecting its rights and privileges against the designs of a medical monopoly poised to guide legislative changes for its own benefit and to curb the "cult" of osteopathy. In a 1939 legislative report to the society, he described his study of legislation pending in several states. It revealed, he wrote, "in almost every instance . . . from six to as many as twenty-five or thirty measures where the rights and privileges of the profession" were involved. Osteopathy's "chief handicap," he asserted, "is the lack of understanding of the members of the profession at large and their seeming indifference or ignorance of the acuteness and seriousness of the problem."[48]

Profile

James O. Watson, D.O.

James O. Watson, D.O., was the stuff of legend. As chairman of the OOA Legislative Committee; as a member of the State Medical Board from 1943 to 1972; as a founder of Doctors Hospital of Columbus, where he served as chief of surgery from 1940 to 1963—no one has done more for osteopathic medicine in Ohio.

James Orion Watson (1901–1985) was born in Fairfield County and grew up on a farm near Thurston, Ohio. He attended Bliss College, then worked in real estate before deciding to pursue a career in osteopathic medicine.

James O. Watson, D.O., shown here in 1944, helped the OOA win critical legislative victories, including a landmark Blue Cross bill in 1939 and unlimited practice rights in 1943. *Courtesy of Jane Cunningham.*

Watson was graduated from Kirksville College of Osteopathic Medicine in 1926, following which he worked as a family physician in a working-class neighborhood of Columbus for several years before serving a residency in surgery at Columbus Radium Hospital. In 1940, Watson, with two other physicians, purchased controlling interest in the hospital and proceeded to transform it into one of the state's premier osteopathic hospitals.

Throughout a long and distinguished career, Watson was active in both state and national professional affairs. As chairman of the OOA Legislative Committee for almost four decades, he fought doggedly to obtain equal practice rights and recognition for Ohio's osteopathic physicians. Recalling these battles at the close of his career, Watson observed, "The osteopathic profession always had to fight [its] way in the legislature for recognition, and we had to win it a piece at a time."[1] Watson did "win it a piece at a time"—from osteopathic inclusion in the first Blue Cross and Blue Shield insurance plans to the landmark revision of the Medical Practice Act in 1943 giving osteopathic physicians equal practice rights with allopathic practitioners. That year, Gov. John W. Bricker appointed Watson to a four-year term as the first osteopathic member of the State Medical Board, a position to which he was reappointed seven times.

Of distinguished mien, with thick white hair and a penchant for bow ties, Watson looked, acted, and spoke like the consummate physician. With his serious demeanor and excellent communication skills, he commanded attention. As a trustee of the American Osteopathic Association, Watson served as chairman of its legislation and public relations departments. He was a Fellow of the American College of Osteopathic Surgeons, which he served as president in 1952–53.

Watson's many awards testify not only to his distinguished leadership, but also to the high esteem in which he was held by his colleagues. In 1954, he received the first "D.O. of the Year" award from the OOA for service to his community and profession. In 1955, he received the AOA Distinguished Service Certificate for his work in osteopathic legislation and organization. Watson was the recipient of the Orel F. Martin Medal, the highest honor of the American College of Osteopathic Surgeons. He was selected by the AOA as the A. T. Still Memorial Lecturer and by the American Osteopathic College of Radiology as the Floyd J. Trenery Memorial Lecturer. At the 1972 OOA convention, held in Cleveland, Watson was named Honorary President of the OOA—the first and only such award ever bestowed. At the same meeting, the annual J. O. Watson, D.O., Memorial Lecture

Drs. James O. Watson (*left*) and Ralph S. Licklider (*right*), shown here in front of Doctors, the hospital they founded, were among the patriarchs of the Ohio profession. *Courtesy of Ohio Osteopathic Association.*

was established to honor his ideals and example.

In 1978, the board of trustees of Doctors Hospital established the James O. Watson Award of Excellence to recognize individuals who have demonstrated outstanding service to the hospital, the osteopathic profession, and the community. Watson himself was the first recipient.

Watson received two honorary degrees of Doctor of Science: from Kirksville, his alma mater, in 1954, and from the Des Moines Still College of Osteopathy and Surgery in 1956. At the Des Moines ceremony, he delivered the commencement address. "Your mind is the master key to the good life," he told the graduates, emphasizing that new doctors should write regularly and learn to estimate and criticize their own work. "Allow time for daily reading. Read newspapers with discrimination; give more time to books. Use the free public libraries, form a library of your own. Spend at least as much for books as you do for movies."[2] Perhaps Watson's greatest contribution to osteopathic medicine was his ability to analyze problems and provide timely advice. "His strong, steady demeanor together with a thoughtful and deliberate approach would provide the profession with some of its best thinking," former OOA Executive Director Richard Sims said in 1999.[3]

Watson died at Doctors Hospital on December 7, 1985, of an apparent heart attack. He was eighty-four. The *Buckeye Osteopathic Physician* eulogized him this way: "Dr. Watson was, simply, a great man who accomplished great things for osteopathic medicine in Ohio."[4]

1. Unpublished interview with J. O. Watson, D.O., files of the OOA, Columbus, Ohio, 6–7.
2. Watson's address was reported in *BOP,* November 1956, 3.
3. "J. O. Watson, D.O., Memorial Lecture," June 24, 1999, OOA, 8.
4. *BOP,* January 1986, 5.

It was increasingly evident that the profession's needs had grown beyond the capacity of its largely volunteer staff and meager budget.[49] In June 1938, after a brief tenure, *Buckeye* editor Dr. Raymond Keesecker and associate editor Dr. Leonard C. Nagel resigned their journalistic posts, citing the need for an executive secretary to take responsibility for the publication.[50] Dr. Meryl A. Prudden briefly took the reins, but the Ohio Society of Osteopathic Physicians and Surgeons seemed to be at a turning point. In October 1939, the *Buckeye* reported that membership was "the lowest it has been in years."[51]

MOVING FORWARD

More people in Ohio know that the Osteopathic Physician is slowly
but surely becoming a fighting organization that will not stand to be
run over and is demanding and will get its just dues.

—*M. A. Prudden, D.O., 1937*

NINETEEN THIRTY-NINE MARKED A TURNING POINT for osteopathic
medicine in Ohio. Under the aggressive leadership of President Ralph S.
Licklider, D.O. (1897–1988), the Ohio Society of Osteopathic Physicians
and Surgeons regained its footing. That year, the board of trustees voted to
raise annual dues from $12 to $25 and to take what Licklider called "one of
the greatest steps forward in the guidance and protection of Osteopathy
that the Association has made in many years." Meeting at the Deshler-
Wallick Hotel in Columbus, trustees voted to establish a new central office
in Columbus and, upon the recommendation of a search committee headed
by Dr. James Watson, to hire William S. Konold as executive secretary.[1]
From offices at 50 East Broad Street, Konold would infuse the professional
association with new energy and a pro-
gressive spirit. Writing in one of his first
issues as editor of the *Buckeye Osteopathic
Physician*, Konold declared his aim: "We
are striving to make Ohio the finest state in
the union for the osteopathic school of
medicine."[2]

Ralph S. Licklider, D.O., of Columbus, served two
terms as president of the Ohio Society of Osteopathic Physicians and Surgeons, from 1939 until
1941. Under his leadership, the organization moved
forward with new initiatives and hired its first paid
executive. *Courtesy of Ohio Osteopathic Association.*

Profile

William S. Konold

"For personal reasons," Bill Konold told a meeting of the Society of Divisional Secretaries upon retiring as executive secretary of the Ohio Osteopathic Association, "I was dedicated to the cause of osteopathic medicine. It was a respected minority group with many friends everywhere but in the right places. The majority group, strong and powerful, held the D.O. by the neck and squeezed."[1]

No layman has done more to advance osteopathic medicine in Ohio than Bill Konold. As executive secretary of the OOA for nearly three decades—from 1939 until 1968—Konold raised the professional standing of osteopathic medicine and helped secure the revision of the Ohio Medical Practice Act, giving the state's D.O.'s full and unlimited practice rights. As a longtime consultant and hospital administrator—"kind of like an itinerant preacher," was how he once put it—Konold helped osteopathic groups in Ohio and elsewhere start new hospitals during the robust period of federally assisted expansion following World War II.

Born in Pittsburgh, William Saints Konold (1898–1977) was reared both in Pittsburgh and Warren, Ohio.[2] During World War I, he enlisted in the Marine Corps, where he served on the battleship *Arizona*. Military service led to his later active involvement with the American Legion, which he served both as state commander (1934) and legislative chairman for twenty-five years. Following the war, Konold attended the University of Illinois, then joined the Warren Tool & Forge Company, a family business, until it closed during the Depression. He worked briefly for the Republic Steel Corporation before starting his own business consulting firm, William S. Konold & Associates. In December 1939 he was hired as the first executive secretary of the Ohio Society of Osteopathic Physicians and Surgeons. (The organization was renamed the Ohio Osteopathic Association of Physicians and Surgeons the following year.) Konold himself dictated the terms of his service, which gave him the latitude to work for other clients at the same time. Thus, when a group of Columbus osteopathic physicians approached him for help in taking over the old Radium Hospital, Konold agreed, becoming the administrator of Doctors Hospital and gradually building it from a 12-bed facility to a 525-bed modern hospital with two locations.

Konold wore many hats—business consultant, hospital administrator, trade association manager, public relations advisor, editor—managing numerous accounts simultaneously during his long career. Besides Doctors Hospital, Konold at various times served as administrator or consultant to Grandview (Dayton), Parkview (Toledo), Mahoning Valley Green Cross (Warren), and

William S. Konold (1898–1977) served as executive secretary of the Ohio Osteopathic Association from 1939 until 1968. During his long tenure, the square-jawed, no-nonsense executive raised the standing of osteopathic medicine in Ohio and nationally. *Courtesy of Mary Jane Carroll.*

Otto C. Epp Memorial (Cincinnati) hospitals. He also counted, among other clients, the Ohio State Veterinary Medical Association and the Ohio State Restaurant Association. As secretary-treasurer of the American Osteopathic Hospital Association from 1940 to 1952—a period when osteopathic hospitals were sprouting nationwide—he guided the fledgling service organization, publishing the first issue of *Osteopathic Hospitals* and helping to obtain recognition for osteopathic institutions by the growing Blue Cross plan. A bespectacled, mustachioed man with a prominent chin, Konold was both an engaging conversationalist and a tough, no-nonsense businessman. "He was a great delegator," is how his daughter, Mary Jane Carroll, explains his ability to juggle the demands of so many clients, while a profile in the *Columbus Dispatch* described him as "an '18-hour-a-day' addict of the science of management."[3] Konold, one colleague recalled, was "the kind of fellow who, when he got into something, got into it with both feet. . . . He was always available and always of great help to those who had questions or needed a conference."[4]

A 1964 resolution recognizing Konold for twenty-five years of service as OOA executive secre-

tary described a man who had, "through his leadership, counsel and wisdom, conceived or been a part of all facets of the development of this profession on a state and national level."[5] In 1968, the American Osteopathic Association honored Konold with the Distinguished Service Certificate—its highest award—for outstanding contributions to osteopathic organization and development.

Following his retirement as OOA executive secretary in 1968, Konold continued as chief executive officer of Doctors Hospital in Columbus until his death there on December 1, 1977.

1. "Twenty-nine Years of Loyal Opposition," address to the Society of Divisional Secretaries, annual meeting of the AOA House of Delegates, Chicago, Ill., July 21–23, 1968.

2. Konold's biography is based largely on a file of personal materials, including Konold's c.v., owned by his daughter, Mary Jane Carroll. See also "Konold Retires," *BOP*, July 1968, 9–10; and "W. S. Konold, Former OOA Executive, Dies," *BOP*, January 1978, 12.

3. "Humor, Pride, Variety Spice Hospital Director's Career," *Columbus Dispatch*, March 24, 1974, 33A.

4. Reminiscence of Robert P. Chapman, secretary of the Society of Divisional Secretaries, in "William S. Konold Receives DSC," *BOP*, October 1968, 20.

5. *BOP*, June 1964, 3.

Konold's sure hand was immediately apparent. "Where did I fit into the picture?" he would later write. "We analyzed the situation, past and present, and charted an organized course of action. . . . I never dictated policy or attempted to become a power in the organization. I did what I was supposed to do in guiding the development of policy. Once policy was established, I drove and drove hard to see that it was carried out."[3] With President Licklider and others, Konold met with officers and members of each of the seven district academies. He advanced programs to enhance the annual convention, including scientific and commercial exhibits, and social events for doctors and their spouses. He advocated on behalf of requiring physicians to register and to participate in two days of postgraduate training each year (a requirement adopted in 1941). And, beginning in February 1940, he expanded the monthly *Buckeye Osteopathic Physician*—the voice of the profession—from eight to twelve pages, adding features on Ohio's osteopathic hospitals and clinics, articles by new doctors and longtime practitioners, and a section on college news. Konold also launched a vigorous drive to increase membership. Noting that 255 men and women were

William S. Konold (*right*) with Michael Ladd, D.O., at an OOA convention in Toledo, date unknown. Konold quickly reinvigorated a struggling organization. *Photo by Hauger & Dorf, courtesy of Ohio Osteopathic Association.*

William S. Konold, hired in December 1939 as the first executive secretary of the Ohio Society of Osteopathic Physicians and Surgeons, scrawled the terms of his contract on hotel stationery. *Courtesy of Ohio Osteopathic Association.*

carrying the load for the state's 464 practicing osteopathic physicians and surgeons, he asked *Buckeye* readers, "Is this equitable and fair?"[4]

At Konold's urging, the board of trustees voted to create a seven-member grievance board for the purpose of disciplining the state's practicing osteopathic physicians. Chaired by John Herbert Brice Scott, D.O., who had practiced in Columbus since 1907 and served as a member of the state Osteopathic Examining Committee, the board was viewed as a first step in raising the standard of osteopathic practice in Ohio. According to the *Buckeye*, the "great majority" of the state's D.O.'s "are sincere to the principles of their profession and the State Society demands that these members not be embarrassed by the unethical practice of a careless minority."[5] In May 1940, the society voted to amend its articles of incorporation, changing its name to the Ohio Osteopathic Association of Physicians and Surgeons (hereafter OOA).[6] Initiatives to boost the image and standing of the profession paid off as membership gradually increased, reaching a new all-time high of 313 in December 1940. The following month, the OOA board of trustees unanimously approved a three-year extension of Bill Konold's contract, sweetening it with a $50-per-month increase.[7] The 44th annual meeting, in May

William S. Konold (*left*) chats with James O. Watson, D.O. (*second from left*), and Domenic J. Aveni, D.O. (*right*). The other individual is unidentified. *Photo by J. Conner Howell & Associates, courtesy of Ohio Osteopathic Association.*

1941, broke all records, with 265 physicians and surgeons and 90 members of the newly organized Auxiliary registered.[8]

Under President Licklider, the board of trustees gave strong support to the legislative committee, which had been invigorated with the appointment, in 1935, of Dr. James Watson as chairman. Watson soon laid out a plan of action at a special meeting of the board called to discuss the legislative situation. He urged trustees to return to their districts and raise an amount equal to $5 per member, the money to be earmarked for the campaign to win equal practice rights and osteopathic representation on the state medical board.[9] The temporary infusion of funds permitted the hiring of an experienced lobbyist, Walter Hamer, who began work in the spring of 1938.[10]

By 1937, twenty-six state legislatures had agreed to extend to D.O.'s the same privileges enjoyed by M.D.'s.[11] But Ohio as yet had been unable to correct its "osteopathic problem": the education of osteopathic physicians long since had evolved to include pharmacology and other therapeutic practices that the 1902 Ohio law forbade them to use in the care of the sick. Writing in the *Buckeye* in 1935, legislative director Mac Hulett described the plight of the Ohio osteopathic physician:

> Osteopathic treatment is fundamentally mechanical, based on adjustments of anatomical structure. But the physicians of that school must use a drug antidote for poison. The law says he shall not.

> The osteopath as a family physician is frequently called upon to vaccinate. When he renders that service, he violates the law by administering a preparation classed as a drug. . . .
>
> If his patient is bitten by a rabid dog, he is denied the right to administer anti-rabies serum; or if attacked by diphtheria, he dare not administer anti-toxin. If in an accident an open wound is infected with tetanus, he cannot, in an effort to save a life, administer tetanus anti-toxin.
>
> Surgery, a part of the osteopathic physician's work, in which he takes the same examination exacted of other physicians, demands a few agents of this kind. But the osteopathic physician commits a felony nearly every time he operates, in order that his patient's life may be spared, or that he may not suffer needless pain.

Because of Ohio's statutory handicaps, Hulett contended, many osteopathic physicians were leaving the state. "The time has long since arrived," he wrote, "when these swaddling clothes with which [the D.O.] was hampered away back over thirty years ago should be laid aside, so that he may keep pace with the modern progress in the healing art. The M.D. has progressed without limitation during this period and the practice of medicine is not at all what it was. But the D.O. has been forced by law to remain as he was, and he must so remain as long as the law remains as it is."[12]

In 1939, the profession marked an important victory with the landmark Blue Cross bill. As written, the bill regulating hospitalization insurance plans provided that subscribers would receive benefits in accredited hospitals "staffed by physicians with the M.D. degree." Watson's efforts to eliminate this blatant discrimination initially proved futile, despite help from a young Marietta attorney, C. William O'Neill, then serving his first term as a representative in the General Assembly.[13] Then, working with Walter Hamer and Paul Riemann, administrator of the Marietta Osteopathic Hospital, Watson drafted a simple amendment. It read: "All service plan contracts issued by such corporation under the provisions of this act shall provide that the subscriber, under such contract, may select for hospital care any non-profit hospital in the State of Ohio and thereupon be entitled to and accorded all benefits contained in the service plan contract."[14] Rep. Lody Huml, Democrat of Cuyahoga County, a longtime friend of osteopathy, introduced the amendment in the House, where, after heated debate, it was approved by a substantial margin. Senate approval followed, and the bill was signed into law. Watson would later describe the amendment as a "Hail Mary pass" that caught the opposition off guard and resulted in a huge victory for osteopathic medicine.[15]

While the profession had prevailed in the Blue Cross contest, the 1936 Bricker ruling restricting the right of osteopathic physicians to administer anesthetics and antiseptics hung like a shadow over the profession and emboldened its opponents. In March 1940, the Darke County Board of Elections, on advice of the Secretary of State, refused to accept petitions presented by Clarence D. Kester, D.O., in support of his candidacy for the office of county coroner. When Ohio Attorney General Thomas J. Herbert ruled that "an osteopathic physician is not a licensed physician" within the meaning of the law establishing qualifications for the office, legal counsel Clarence Corkwell filed a mandamus proceeding in the Ohio Supreme Court. On April 24, 1940, the court ordered the Ohio Secretary of State to place D.O. candidates for coroner on the ballot, declaring, "To rule an osteopathic physician was not a physician would be to say a shepherd dog was not a dog." (Kester was elected, as was A. W. Larrick, D.O., serving Guernsey County.)[16] The same year, the Blind Commission of Ohio rejected examinations by osteopathic physicians—"on the theory that we are not physicians skilled in the knowledge of the eye," fumed Corkwell—and the Erie County prosecutor advised county commissioners that federal and state law prohibited an osteopathic physician from serving as health commissioner. In each case, the vigilant Corkwell stood ready to file mandamus proceedings "to establish our undisputed rights."[17] Minutes of the board of trustees show that he occasionally was also called upon to police the profession, in one case visiting two doctors in Marion who had issued an advertising card "that was anything but ethical."[18]

With America's entry into World War II, OOA President Donald V. Hampton, D.O., of Cleveland, mobilized the state's osteopathic physicians to assist the war effort. Although D.O.'s were excluded from serving in the medical corps, they nevertheless would play "an integral and important part" in the conflict, Hampton wrote in the *Buckeye.* "Many doctors will be called to serve with the armed forces, leaving a greater burden on those left to care for civilian health. . . . Another important service we will render is in civilian defense units. . . . We are volunteering our services 100% to the Civilian Defense organization."[19] Within months of America's entry into the war, Hampton reported that inventories of osteopathic facilities had been forwarded to civil defense authorities and emergency medical field units were being organized in connection with each of the state's osteopathic hospitals.

"EXTRA": Osteopathic physicians rejoiced in 1940 when the Ohio Supreme Court overruled Ohio Attorney General Thomas J. Herbert, who had declared that "an osteopathic physician is not a licensed physician" within the meaning of the law establishing qualifications for coroner. The court declared: "To rule an osteopathic physician was not a physician would be to say a shepherd dog was not a dog." *Courtesy of Ohio Osteopathic Association.*

PRESS TIME

"**EXTRA**"

Osteopathic Physicians & Surgeons are licensed physicians, rules Ohio Supreme Court!

ATTORNEY GEN. OVER-RULED

Full details at the Convention

The Auxiliary to the Ohio Osteopathic Association

Since 1939, members of the Auxiliary to the Ohio Osteopathic Association have donated their time, talent, and money to support osteopathic causes. Comprised of doctors' spouses and friends of osteopathy, the AOOA had its origin at the 1938 state convention in Marietta. There, a group of women interested in the osteopathic profession gathered to hear Helen Conard Keesecker, the wife of Raymond P. Keesecker, D.O. of Cleveland, explain the need for such an organization. Although several district auxiliaries already were working on behalf of osteopathic causes, as yet no attempt had been made to form a statewide organization. The women met again in July 1938 during the AOA convention at the Netherland Plaza Hotel in Cincinnati, where they elected officers. "The women," Ouida S. LaRue later reported, "feel it is time for them to perfect an organization . . . that will be of definite help for the education of the public to the benefits and advantages of the science of osteopathy. The real object of this organization is to disseminate a practical knowledge of osteopathic health service and to assist in the advancement of such service."[1] The first officers of the Ohio Women's Osteopathic Association (later renamed the Auxiliary to the Ohio Osteopathic Association of Physicians and Surgeons) were Mrs. Charles M. LaRue, Columbus, president; Mrs. H. L. Benedict, Marietta, vice president; Mrs. E. C. White, Warren, secretary; and Mrs. Richard A. Sheppard, Cleveland, treasurer.

In November 1938, association members met again in Columbus to draft a constitution and bylaws. Forty women were present for the first regular meeting on May 6, 1939, at the Deshler-

Kay Licklider (Mrs. Ralph S. Licklider) was an active Auxiliary member from its inception, drafting the original constitution and bylaws and later writing a brief history of the organization. *Courtesy of Ohio Osteopathic Association.*

When ladies wore hats: Auxiliary annual meeting held in conjunction with the OOA state convention, Columbus, 1949. *Courtesy of Ohio Osteopathic Association.*

Auxiliary officers, 1949–50. *Left to right:* Mrs. T. C. Hobbs, Delaware, president-elect and reporter, AOOA Bulletin; Mrs. Walter Kelly, Marietta, first vice president; Mrs. W. H. Nicholson, Cambridge, treasurer; Mrs. H. R. Hunter, Akron, president; Mrs. Robert E. Sowers, Warren, secretary; Mrs. William E. Reese, Toledo, historian; and Mrs. D. M. Stringley, Dayton, second vice president. *Courtesy of Ohio Osteopathic Association.*

Wallick Hotel in Columbus, when each of seven district auxiliaries was asked to sponsor a fund-raising project to help finance the Ohio society's new central office. By the end of the year, $1,200 had been turned over to help pay the salaries of the new staff. In addition to fund-raising, district auxiliaries placed osteopathic magazines in public and school libraries and hosted programs to familiarize the public with osteopathic principles. "Our Association is young and growing," Ouida LaRue reported in the *Buckeye,* "but we want to be known as a definite and permanent group of women who are determined to work to educate the Public to the knowledge and benefits of the Science of Osteopathy."[2]

By 1943, Auxiliary members were supporting the state's fledgling osteopathic hospitals—rolling bandages, sewing and knitting, serving as nurses' aides, and buying equipment. Some placed osteopathic literature in college and high school libraries, while others helped in public bond sales. Political action, too, became part of the Auxiliary's work when President Catherine L. Licklider urged members to write to President Roosevelt and to their congressmen to protest the failure of the Surgeon General of the Navy to commission osteopathic physicians and surgeons as medical officers.[3] As the number of osteopathic hospitals in Ohio increased, the Auxiliary played an increasing role in their support. By 1945, 219 Auxiliary members offered their services to Ohio's seven osteopathic hospitals. In her president's message that year, Mildred Clybourne emphasized the importance of osteopathic hospitals "to the continuance and progress of our profession," calling them the "'display windows' of our work."[4]

For six decades, the Auxiliary has promoted and strengthened the osteopathic profession in Ohio—through direct services to hospitals, recognition of hospital volunteers, and fund-raising to finance scholarships, student loans, and research. The AOOA has also sponsored special service projects, such as the distribution of child safety activity books and the recent Yellow Ribbon Program to increase public awareness of the epidemic of teenage suicide. Beginning in the 1980s, Auxiliary members also assisted with proctoring students taking the National Board of Osteopathic Medical Examiners tests. "I would like to think," Auxiliary President Ginny Settembrini wrote in 1988, "that there are 10,000 additional people around the state who are more aware of the osteopathic profession because of our efforts."[5]

1. *BOP,* October 1938, 1.
2. *BOP,* February 1940, 3.
3. *BOP,* November 1943, 6.
4. *BOP,* November 1945, 6.
5. *BOP,* July 1988, 13.

In 1942, the American Osteopathic Association moved its annual convention from Los Angeles to Chicago; the special trains that were to have carried conventioneers and their spouses west could no longer be spared. Subsequent AOA meetings were cancelled for the duration of the war. Ohio D.O.'s continued to convene through 1944, but the 1945 annual meeting and a planned two-day "refresher" course in February that year were cancelled in cooperation with the government's effort to curtail meetings of large groups. Despite wartime, the 1942 convention, at the Deshler-Wallick Hotel in Columbus, broke all records, with 401 doctors, interns, students, and guests attending.

War had a salutary impact on the osteopathic profession: as M.D.'s left to serve in medical units of the armed forces, many civilians at home turned to osteopathic physicians for their medical care. They soon learned that D.O.'s were competent professionals capable of taking care of all their medical needs.[20] Ohio D.O.'s, meanwhile, continued to press their case for equal practice rights. In 1941, the OOA sponsored House Bill 329, introduced by Rep. Lody Huml, providing for representation of the osteopathic school of medicine on the state medical board. "This measure," the *Buckeye* declared, "is sound in every respect. The Ohio State Medical Board is

---

Ohio Women's Osteopathic Association

Before the Auxiliary to the Ohio Osteopathic Association was organized, another group brought together Ohio's women osteopathic physicians and their supporters for the purpose of educating the public about osteopathy. The Ohio Women's Osteopathic Association, a branch of the Osteopathic Women's National Association (OWNA), founded in 1920, provided Ohio's women D.O.'s with opportunities for social contact and to work together on programs promoting osteopathic education and the health and welfare of women and children. Associate membership was open to women friends of osteopathy and women related to osteopathic physicians. Officers of the group in 1923 were Alice P. Bauer, D.O., Delaware, president; Katherine M. Scott, D.O., Columbus, vice president; Carrie E. Hutchison, D.O., Dayton, secretary; Grace Purdum Plude, D.O., Cleveland, treasurer; Margaret E. Wilson, D.O., Sidney, auditor; and Charlotte Weaver, D.O., Akron, press representative. The

Ohio branch of OWNA was among the strongest. By 1934, the group had established a scholarship loan fund "for worthy Ohio women students in osteopathic colleges" and had assisted several students.

In 1934, Helen Marshall Giddings, D.O., of Cleveland, then serving as president of OWNA, presciently wrote, "In the future osteopathy must include in its program a definite plan of encouragement and aid to the lay women's activities. The countless women who are wives and families of osteopathic physicians, and friends of osteopathy, can be united into an army of workers who will—if given certain definite tasks—remove many of the hindrances to osteopathic progress."[1] The Ohio Women's Osteopathic Association disbanded in the 1940s, but its history testifies to the strong presence of women professionals in the early years of osteopathic medicine in Ohio.

1. "O.W.N.A.," *Buckeye Osteopath,* September 1934, 6.

William Konold was the self-styled "Dean" on the cover of the December 1944 *Buckeye. Courtesy of Ohio Osteopathic Association.*

the governing body of all physicians and surgeons regardless of their school of medicine. They conduct the examinations and issue the certificates to practice. The educational requirements for all schools of medicine are now the same. Therefore, it is only reasonable and just that the osteopathic school of medicine should be represented on this governing board."[21] However sound, the measure was defeated in the House Health Committee by a vote of five to three.[22]

Representing some five hundred osteopathic physicians, the OOA was fighting an uphill battle against nine thousand medical doctors represented by the AMA, a wealthy and powerful interest group that had long sought to undermine osteopathy by lobbying against improved osteopathic legislation. Now, the OOA used the obnoxious designs of organized medicine to push for increased membership and financial support for its legislative efforts. In 1943, in the midst of war, the OOA achieved the greatest legislative victory in the state's osteopathic history: the passage of Substitute House Bill 112, giving Ohio D.O.'s full and unlimited practice rights, and providing for osteopathic representation on the state medical board.

Looking back forty years later, Dr. James Watson summarized the path to victory in his characteristic low-key manner:

> All of those licensed in Ohio up to that point were examined first by what was called an "Osteopathic Examining Committee" in several subjects and then, if they were successful, they were examined in additional subjects . . . taking the same examination in those five subjects taken by all other candidates. If they were successful, at this point they were then licensed to practice osteopathic medicine & surgery, but not to administer or prescribe drugs except anesthetics and antiseptics. From year to year, we [had] made attempts in the Legislature to obtain revisions of the Medical Practice Act but without success. . . . In the fall of 1939, Mr. W. S. Konold came on as Secretary of the Ohio Osteopathic Association and early on, we recognized that we would need a new lawyer and a different kind of lawyer with knowledge and skill politically as well as in the law. . . . Charles J. Chastang came on board.[23]

Chastang (1906–1994), a young associate in the law office of Ed Schorr, chairman of the Ohio Republican Party, proved an important addition to the OOA team. Representatives of the Ohio State Medical Association (OSMA) and the Ohio Osteopathic Association (for OOA, they were Watson, Chastang, and Konold) hammered out the final bill. Watson explained how the OOA was able to overcome the medical society's historic opposition to osteopathic representation on the state medical board. "The Medical Board at that time was made up of seven physicians and further provided, taking into account that there were allopathic physicians, homeopathic physicians and eclectic physicians, that no one school of practice would have a majority. . . . Obviously, as time went on, this was commencing to be disadvantageous to the medical association in view of the diminishing numbers of homeopathic physicians and even more diminishing numbers of eclectic physicians. Thus, the OSMA had a stake [in changing the makeup of the board]."[24]

As a young attorney serving as legal counsel to the OOA, Charles Chastang played a key role in securing unlimited practice rights for Ohio's osteopathic physicians. *Courtesy of Ruth Chastang.*

The revised Ohio Medical Practice Act, which became effective July 30, 1943, provided for a state medical board consisting of eight physicians—seven M.D.'s and one D.O. Gov. John W. Bricker appointed Dr. James Watson as the first osteopathic member of the state medical board, a position he would hold for nearly three decades.[25] As the osteopathic member of the board, Watson prepared the examination questions in osteopathic materia medica and principles and practice of osteopathic medicine.[26] To obtain full and unlimited practice rights, D.O.'s already practicing in Ohio could obtain thirty-six weeks of special education and thereby become eligible to take the examinations. A small group of doctors—"perhaps close to 50," Watson later estimated—determined that they would not qualify and chose not to sit for the examinations; they were permitted to continue as limited practitioners. However, "the majority . . . could qualify and did write the special examination required in the four subjects, which, if successful, led to their full and unlimited licensure."[27] The revised law also provided for the establishment of an education committee comprised of one member of the state medical board having an M.D. degree, one member having a D.O. degree, and the state superintendent of education. The committee (including Watson, representing D.O.'s) visited and inspected the six osteopathic colleges to determine "that the colleges were qualified and did train their graduates in a satisfactory manner." All were approved, making their graduates eligible for state board examinations in Ohio.

Ohio's D.O.'s had finally achieved full and unlimited practice rights and representation on the state medical board. It was a giant step forward for the profession and for the Ohio Osteopathic Association. Both were now well positioned to ride the coming wave of postwar growth and prosperity.

4

POSTWAR STRIDES AND SETBACKS

The child Osteopathy is now a strong individual, commanding the
respect of scientific men everywhere.

—*Walter H. Siehl, D.O., 1948*

AT THE CLOSE OF WORLD WAR II, Ohio counted slightly more than five
hundred osteopathic physicians. Ninety-seven percent were dues-paying
members of the Ohio Osteopathic Association, compared with just 67 percent two years earlier.[1] As advertisements in the *Buckeye* foretold a new
postwar world that would enjoy atomic energy for peacetime uses, all-
weather aviation, and a new national highway system, the OOA, buoyed
by the passage of a new Medical Practice Act, enjoyed two decades of sustained progress under the leadership of Executive Secretary Bill Konold.
The postwar period saw the construction of new osteopathic hospitals and
the expansion of existing hospitals, the construction of a new OOA central
office, and solid gains in the status and recognition of the osteopathic profession. While prejudice did not disappear, it waned markedly. In 1956,
President Dwight D. Eisenhower signed Public Law 763, permitting D.O.'s
to be commissioned in the medical corps of the armed services. In 1961, the
AMA, which had long deemed M.D.–D.O. professional contact to be "unethical," gave its constituent societies the right to determine whether or not
its members could voluntarily associate with osteopathic practitioners on a
professional basis. And in 1973, the last of the "limited practice" states,
Mississippi, licensed D.O.'s for the full practice of osteopathic medicine and
surgery. Despite these advancements, the profession continued to grapple
with questions of image as it worked to educate a public still prone to con-
fusing the osteopathic physician with the chiropractor and the orthopedist.

In 1945, the Ohio Osteopathic Association adopted a new constitution
and bylaws creating a House of Delegates proportionate to the members in

June B. Day, William Konold's longtime executive assistant and bookkeeper, with orthopedic surgeon Harold E. Clybourne, D.O., 1947. *Courtesy of Ohio Osteopathic Association.*

its geographic districts.[2] (Prior to 1945, OOA officers had been chosen by members at the annual convention from a slate proposed by the board of trustees.) Two years later, it unveiled a "revamped and enlarged" professional magazine. The redesigned *Buckeye Osteopathic Physician* publicized the activities of its now fifteen district academies and encouraged doctors to become involved in community service and local public health programs as a means of gaining positive public exposure for the profession and raising the profile of osteopathic medicine.

At the end of World War II, osteopathic medicine in Ohio was just shy of the half-century mark. Speaking at the president's banquet in May 1947, OOA Past President Charles Rauch, D.O., of Logan, warned that, having won equal practice rights, Ohio's osteopathic physicians could not afford to develop a false sense of security. He identified three "currents" that might "easily throw us from our course": failure to support the OOA, failure to support the osteopathic colleges, and failure to counter the challenge from those who believed "we should allow allopathic medicine to take us into her fold." On the last point, Rauch was adamant, saying, "We can no more think of surrendering osteopathy to be engulfed by allopathy than we can think of surrendering the freedom of our democracy to a dictatorship."[3]

With 454 members of the American Osteopathic Association, Ohio in 1949 ranked fifth in the standings by state, after California (with 1,166), Michigan (749), Missouri (686), and Pennsylvania (648).[4] Thanks in part to effective vocational guidance programs organized by the district academies, the state was also a leader in contributing students to the nation's six osteopathic colleges. In 1949, 135 Ohioans were enrolled: 48 at Kirksville, 31 at Chicago, 24 at Kansas City, 21 at Des Moines Still, 6 at Philadelphia, and 5 at California. (Of these, only four were women.)[5] Five years later, OOA President W. Dayton Henceroth, D.O., announced that 77 recent graduates—"a new all-time high"—had taken the state medical board examinations, a tribute to the work of the OOA Physicians Location Committee, which actively recruited new graduates to establish medical practices in

Ohio. "It was inspiring," Henceroth reported, "to walk among these enthu-siastic doctors and welcome them to the state of Ohio in behalf of the most rapidly growing [osteopathic] association in the country."[6]

With no tax support, no university affiliations, and scant outside phi-lanthropy, the profession itself bore increasing responsibility for the up-keep and improvement of its colleges. In 1946, Kirksville President Morris Thompson visited Ohio on behalf of the schools, declaring that the six os-teopathic colleges would need $1.7 million annually in order to expand into "first-class institutions." D.O.'s were asked to contribute to the Osteopathic Progress Fund, established by the AOA in 1943; funds subscribed would be channeled directly to the medical schools. The campaign was given added urgency following the Surgeon General's announcement, in 1959, that the nation faced a serious doctor shortage. That year the OOA, which until then had relied on voluntary contributions to support the colleges, voted to raise annual dues from $75 to $175, with $100 of that amount earmarked for the Osteopathic Progress Fund.[7] In tandem with doctors' support of the colleges, the AOA's Osteopathic Seal program, begun in 1931, gave lay friends of the profession an opportunity to contribute to the advancement of osteopathic medicine. The money raised through purchase of the seals was earmarked for student loans and research. "Remember—a seal on every Christmas card helps tell our profession's story," the *Buckeye* re-minded readers in 1959.[8]

Following World War II, U.S. hospitals enjoyed dynamic growth fu-eled by an exploding birth rate, the expansion of private health insurance, and passage of the Hospital Survey and Construction Act in 1946. Widely known as the Hill-Burton Act after its sponsors, Senators Lister Hill and Harold Burton, the purpose of the legislation was to provide federal grants

Green Cross General Hospi-tal opened in Akron in 1943. *Photo by Jack Lehr, Acme Pic-tures, courtesy of Ohio Osteo-pathic Association.*

to modernize the nation's hospitals, which had become obsolete due to lack of capital investment during the Great Depression and World War II. The program pumped massive amounts of federal money into the construction of new health care facilities. By the close of 1953, Ohio ranked second (behind only Texas) in the number of Hill-Burton projects, having received $85 million and added 4,757 new beds to its inventory.[9]

Osteopathic hospitals grew prodigiously during this period. On the eve of World War II, there were six osteopathic hospitals; by 1965, there were seventeen. The pace was heady, with virtually every issue of the *Buckeye* carrying news of building plans, groundbreakings, or dedications. New osteopathic hospitals opened in Akron in 1943, in Toledo and Warren in 1946, and in Bay Village, a suburb of Cleveland, in 1948. Doctors Hospital in Columbus undertook several major expansion projects during the 1950s, as well as the construction, in 1963, of a "satellite" hospital—Doctors West—in the Columbus suburb of Lincoln Village. Dayton's Grandview Hospital moved to a new location in 1947 and expanded with two major additions, in 1952 and 1958. New osteopathic hospitals opened in Orrville, Sandusky, and Youngstown in 1953; in the southeast Cleveland suburb of Warrensville Heights in 1957; in Warren in 1958; in Madison (Lake County) in 1959; in Cincinnati in

Doctors observe surgery at Green Cross General Hospital, Akron, date unknown. *Courtesy of Ohio Osteopathic Association.*

Doctors Hospital, Columbus, greatly expanded after World War II, adding two new wings in the 1950s. *Photo by J. Conner Howell & Associates, courtesy of Ohio Osteopathic Association.*

Opened in 1963 in suburban Columbus, the 112-bed Doctors West, a unit of Doctors Hospital, pioneered the concept of the "satellite" hospital. *Courtesy of Ohio Osteopathic Association.*

1960; and in Richmond Heights (Cuyahoga County) in 1961. Doctors Hospital of Stark County, located between Canton and Massillon, and Selby General Hospital in Marietta—the last independent osteopathic hospitals to be built—opened in 1963 and 1965, respectively. In the midst of this building boom, Bill Konold offered statistics to demonstrate the osteopathic profession's growing contribution to the health care needs of Ohio's citizens: in 1956, the state's osteopathic hospitals admitted almost 33,000 Ohioans, brought over 6,000 newborns into the world, performed 17,000 surgeries, and cared for over 8,000 medical cases. Ohio was becoming one of the largest osteopathic states, behind only California, Michigan, and Pennsylvania.[10]

In 1956, OOA President John W. Hayes, D.O., cut the ribbon to open the new central offices of the Ohio Osteopathic Association. "Spacious,

Photo by Mound Photographers, courtesy of Ohio Osteopathic Association.

Photo by Charles R. Brown, courtesy of Ohio Osteopathic Association.

Ohio saw a flurry of hospital construction following World War II, much of it aided by federal money through the Hill-Burton Act of 1946. New osteopathic hospitals opened in (*from top*) Sandusky (1953), Youngstown (1953), Warrensville Heights (1957), and Cincinnati (1960).

Photo by Scope Photographers, courtesy of Ohio Osteopathic Association.

Photo by Howard A. Newman, courtesy of Ohio Osteopathic Association.

comfortable and modern," according to the *Buckeye*, the two-story brick-and-concrete-block building at 53 West Third Avenue, one block east of Doctors Hospital, was erected at a cost of $50,000 and financed by loans from OOA members. Within two years, the debt was retired. "This transaction . . . augurs well for the stability and prestige of our organization," said OOA President Robert L. Thomas, D.O.[11]

In 1961, the profession was jolted when the members of the California Osteopathic Association, the nation's largest divisional society, voted to merge with the California Medical Association and to exchange their D.O. degrees for M.D. certificates. Driving the move, among other factors, was

Courtesy of Ohio Osteopathic Association.

In 1956, President John W. Hayes, D.O., cut the ribbon
to open the new central offices of the Ohio Osteopathic
Association of Physicians and Surgeons at 53 West
Third Avenue, Columbus, one block east of Doctors
Hospital.

*Photos (top and above) by Daniel Firestone. Courtesy of Ohio
Osteopathic Association.*

the inadequate financing of osteopathic education, inadequate postgraduate training, the perceived poor status of the D.O. degree, and osteopathic
physicians' exclusion from group health insurance plans. The California
Medical Association ratified the merger plan, and in a special referendum in
November 1962, California voters approved a proposition removing the
power of the Board of Osteopathic Examiners to license new D.O.'s. The
merger also meant that one of the nation's premier osteopathic colleges, the
College of Osteopathic Physicians and Surgeons in Los Angeles, was now
lost to the profession; it would become an allopathic medical college and
award the M.D. degree to future graduates.[12]

The Sheppards of Bay View

A notorious murder, still unsolved almost a half-century later, has overshadowed the story of one of Ohio's most prominent osteopathic families. Its patriarch, Richard Allen Sheppard, D.O. (1890–1955), was born in Fostoria, Ohio. His father was a Methodist minister; his mother, state president of the W.C.T.U. A younger sister who suffered from rheumatic heart disease—for which conventional medicine offered no treatment—led him to study osteopathic medicine. After graduating from the American School of Osteopathy (Kirksville) in 1911, Richard Sheppard practiced in Fayetteville, North Carolina, and Upper Sandusky, Ohio, before settling in Cleveland in 1923. There, he joined the Roscoe Osteopathic Clinic, later opening the downtown Cleveland Osteopathic Clinic with five other physicians. Sheppard initially specialized in obstetrics/gynecology and later became a general surgeon. He performed surgery at the East 79th Street Hospital in Cleveland—one of the few area hospitals open to D.O.'s—but when that facility was closed during the Depression, the situation led him to found the Cleveland Osteopathic Hospital, one of the first osteopathic hospitals in Ohio. Sheppard was chief of staff of the new hospital, which opened in 1935 in a former private mansion on Euclid Avenue, Cleveland's once-famous Millionaires' Row.

Richard A. Sheppard, D.O., in his office at the Cleveland Osteopathic Hospital, about 1940. *Courtesy of Ohio Osteopathic Association.*

Sheppard enjoyed a distinguished career, twice serving as president of the Ohio Society of Osteopathic Physicians and Surgeons (Ohio Osteopathic Association), in 1923–24 and 1936–37. He also served as the first president of the Ohio Osteopathic Hospital Association, founded in 1939, and as president of both the American Osteopathic Hospital Association and the American College of Osteopathic Surgeons. In 1944, he received an honorary Doctor of Science degree from the Los Angeles College of Osteopathic Physicians and Surgeons (COPS), where he taught in the graduate school at least one month each year until 1953. His wife, Ethel Niles Sheppard, was long active in the Ohio and national osteopathic auxiliaries, serving as president of both groups.

In 1948, Richard Sheppard opened a new hospital in the Cleveland suburb of Bay Village, where he served as chief of staff and practiced alongside his three sons, also surgeons: Richard Niles, Stephen Allen, and Samuel Holmes. "Call Dr. Sheppard—and you'll get a crowd" read the lead of a feature story about the family of surgeons in the *Cleveland News*.[1] Richard (1916–1980), the oldest, was a gynecologist and obstetrician. Stephen (b. 1920) was a urologist. Sam (1923–1970), youngest of the three, was a neurosurgeon. All three were graduates of COPS (Los Angeles).

Sam Sheppard joined the Bay View staff in 1951 following a residency at Los Angeles County General Hospital. He was a rising star and active in his field, presenting papers to both the West Virginia and Colorado osteopathic conventions in 1951. He served as police and fire department surgeon for the communities of Bay Village and Westlake, and in 1953 delivered a talk on "The Neurological Examination" at the OOA convention.[2] The Sheppards constituted a "close-knit unit," Stephen Sheppard wrote in his 1964 book, *My Brother's Keeper*. "We shared each other's homes and problems, working together at the Sheppard Clinic [a satellite facility opened in Fairview Park in 1952] and Bay View Hospital."[3] Then tragedy struck.

Bay View Hospital as it appeared in the early 1950s. *Photo by Deming Photo Service, Cleveland, courtesy of Ohio Osteopathic Association.*

"My father and mother had moved only four months before [the murder] into a big white house on a hill overlooking Lake Erie and immediately east of the hospital. On the other side of the hospital and also overlooking the lake was another white house in which Richard lived with his wife, Dorothy, and their three children. . . . Several miles to the west but on the same street Sam lived in another white lakeshore home with his wife, Marilyn, and their son, Sam, known as Chip." Stephen lived nearby in Rocky River, a suburb just east of Bay Village, with his wife and two children.

On July 4, 1954, Sam and Marilyn were to host a picnic for a group of Ohio surgeons. That morning, Marilyn was found brutally murdered in her bedroom. Days later, Sam Sheppard was charged with the crime. According to Stephen Sheppard, "We never for a moment doubted his innocence, either on the day of the crime, the day of his arrest, the day of his conviction, or on the many gloomy days thereafter when court after court turned him down. His guilt, to us, was simply a physical, as well as psychological, impossibility." The authorities, however, from the start regarded Sam as not only the number one suspect, but as the only suspect; and, in Stephen Sheppard's view, "they believed—or pretended to believe—that the rest of us were his confederates in trying to save him from the con-

sequences of the crime." He would later write of watching "helplessly while my mother and father shriveled up before my eyes in the intense heat generated by the suspicion directed at all of us. It was apparent within a few days to Richard and me that the developing ordeal would be too much for our parents—that to all intents and purposes their lives were over."[4]

Sam Sheppard was quickly convicted of murder in a trial-turned-media-circus. On January 7, 1955, Ethel Sheppard shot herself in the forehead with son Stephen's .38 caliber revolver. Incapacitated by a stroke, her husband Richard died eleven days later. He was sixty-four. At a memorial service held at Bay View Hospital, Raymond P. Keesecker, D.O., editor of the *Journal of the American Osteopathic Association* and a former colleague, said that the elder Sheppard would be remembered as a "builder and the wise and mature healer of the ills of men's bodies."[5]

As the legal appeals dragged on, Stephen Sheppard wrote, "My work and Richard's work increased in volume. . . . It became evident that

The four Sheppard doctors in happier times, in the early 1950s. *Left to right:* father Richard A. with sons Stephen, Richard, and Sam. *Courtesy of Sam Reese Sheppard.*

Following in his father's footsteps, Sam Sheppard, D.O., was a rising star in the osteopathic profession when this photograph was taken, about 1951. *Courtesy of Ohio Osteopathic Association.*

the Sheppard case had had little, if any, effect upon our professional careers or upon the welfare of Bay View Hospital. The hospital was bulging at its seams. Crash plans for expansion of facilities were under way." Then, about May 1963, Stephen wrote, "I was called on the carpet at Bay View Hospital by an old friend who was a member of the hospital's recently organized public relations committee. He said he had been delegated by the committee to discuss the problem of projecting a desirable image of the hospital in the face of harmful publicity. . . . He came to the point with commendable bluntness," suggesting that Stephen and Richard withdraw from the staff and establish their practices elsewhere. "The committee," Stephen was told, "wants to get away from the idea that this is 'The Sheppard Hospital.'"

Stephen Sheppard countered the suggestion head-on.

"Don't you know that long after Richard and I are dead and buried, Bay View will still be known as 'The Sheppard Hospital?'" I asked him. "Let's face it. My father did found this hospital. He and my mother built it. They mortgaged everything they owned in the depths of the depression in order to make this hospital possible. My father paid the bills for years and my mother worked in every department from laundry to dietary when she was needed.

"That's what makes it 'The Sheppard Hospital.'"

Stephen Sheppard continued:

I clicked it off for him—thirty-five beds when my father started; fifty-three beds and eighteen bassinettes [*sic*] in 1954; one hundred and ten beds a year ago, and two new stories just completed that would add seventy beds in another month.

"This we have done during a period when we were subjected to the most damaging sort of publicity imaginable," I said.

I asserted that the only kind of public relations that really counts is the day-to-day care of patients and attention to detail in little things. I said this had enabled us to survive in the last nine years—not only to survive, but to more than treble our hospital capacity.

He asked me if this meant I intended to continue along the lines I had apparently laid down for myself over the last nine years.

"I mean that I shall employ whatever means are at my disposal to state in both public and private at every opportunity and in every way that my brother Sam is innocent and should be released from prison."[6]

In 1968, Stephen Sheppard announced that he would leave the osteopathic profession and move to California to begin a residency in psychiatry at the Napa State Hospital. There, he would practice as an M.D. In a bittersweet farewell to the profession, he wrote of watching "our profession degenerate into a mass of conformists." In striving to become as much like M.D.'s as possible, he wrote, "we have lost some of the magic which our forebears used with skill and insight."[7]

After almost ten years in prison, Sam Sheppard, represented by a then little-known defense attorney named F. Lee Bailey, won a new trial. In a stinging rebuke to the original trial judge and the Cleveland newspapers, U.S. District Court Judge Carl Weinman of Dayton called the 1954 trial a "mockery of justice" and ruled that Sam Sheppard had not received a fair trial. The U.S. Supreme Court subsequently agreed, striking down Sheppard's murder conviction. Cuyahoga County retried Sam Sheppard, and a second jury found him not guilty.

Sam Sheppard applied to rejoin the Ohio Osteopathic Association. In 1968, he briefly joined the staff of Youngstown Osteopathic Hospital, where he was charged in two wrongful-death suits before resigning. Wasted by drink and drugs, Sheppard moved to Gahanna, Ohio, where he died of liver failure on April 6, 1970. He was forty-six.

In January 2000, Sam Sheppard's son, Sam Reese Sheppard, addressed members of the

With his wife Ariane, Sam Sheppard, D.O., looks over the program for a postgraduate seminar sponsored by the Cleveland Academy of Osteopathic Medicine, January 1968. Sheppard ultimately was unable to resume the successful surgical career he had enjoyed prior to his conviction and incarceration. He died in 1970 at age forty-six. *Courtesy of Cleveland Press Collection, Cleveland State University Library.*

Cleveland Academy of Osteopathic Medicine. He was in Cleveland on the eve of a wrongful-imprisonment suit filed on behalf of his father—a legal fight he would lose. Speaking at the Beachwood Marriott on "The Sam Sheppard Case and Its Osteopathic Connection," he recounted the prejudice that osteopathic physicians historically had faced, describing, in particular, the many trials his grandfather, Richard A. Sheppard, had endured in his attempt to build the osteopathic equivalent of the Cleveland Clinic. Sheppard's appearance before a standing-room-only audience of osteopathic physicians—members of a profession that, embarrassed by the notoriety of the Sam Sheppard murder case, had turned its back on the Sheppard family—seemed to bring a tragic story full circle.

In 2001, after painstakingly combing the facts, analyzing once-discarded Cleveland police reports and grand-jury transcripts, and weighing new evidence and the judgments of independent forensic experts, investigative reporter James Neff concluded, "Dr. Sam Sheppard did not kill his wife." Instead, he pinned the blame elsewhere, soberly observing that "an enormous mistake had been made against a family, one that was impossible to undo."[8]

1. The feature article was reprinted under the title "The Sheppards of Bay View Hospital," *BOP*, August 1951, 9.

2. "Briefs on O.O.A. Convention Speakers," *BOP*, April 1953, 11.

3. Stephen Sheppard with Paul Holmes, *My Brother's Keeper* (New York: David McKay, 1964), 8.

4. Sheppard, *My Brother's Keeper*, 46, 57, 59.

5. *BOP*, July 1955, 11. In 1975, Richard A. Sheppard, D.O., posthumously received the Award of Merit from the Cleveland Academy of Osteopathic Medicine for "outstanding contributions to the well-being of the osteopathic profession." Although the *Buckeye Osteopathic Physician* reported the elder Sheppards' deaths, it reported neither the murder of Marilyn Sheppard nor the trial and incarceration of Dr. Sam Sheppard.

6. Sheppard, *My Brother's Keeper*, 220, 288–90.

7. Stephen A. Sheppard, D.O., "Why I Am Leaving Osteopathy," *OP/The Osteopathic Physician*, May 1968, 47.

8. James Neff, *The Wrong Man: The Final Verdict on the Dr. Sam Sheppard Murder Case* (New York: Random House, 2001), 382.

Outside California, the merger seemed only to exacerbate the mutual mistrust between the American Osteopathic Association and the American Medical Association. In his 1961 report to the OOA House of Delegates, Bill Konold analyzed the threat that the California merger posed to the future of osteopathic medicine in Ohio. Mincing no words, he said, "Now that you have achieved and maintained your independent school of medicine, the other school of medicine is showing an interest in eliminating you. . . . In my opinion, they have used California as a pilot state toward this objective." Amalgamation would mean "coming into their scheme of things through the back door"; Ohio osteopathic physicians would become "M.D.'s of questionable education . . . constantly trying to convince the public that [they] are just as good as the others." Konold then outlined a ten-point program, emphasizing the importance of supporting the colleges and strengthening Ohio's seventeen osteopathic hospitals. Physicians, he said, must apply the "osteopathic concept" in their practices. "People want this and it is what distinguishes you from the run-of-the-mill doctor. Use it. It's your future," he told OOA members.[13]

In June 1961, the AMA House of Delegates voted to give each state medical society the right to determine whether or not its members could voluntarily associate with osteopathic practitioners on a professional basis. "The test," the AMA declared, "should now be: Does the individual doctor of osteopathy practice osteopathy, or does he in fact practice a method of healing founded on a scientific basis? If he practices osteopathy, he practices a cult system of healing and all voluntary professional associations with him are unethical. If he bases his practice on the same scientific principles as those adhered to by members of the American Medical Association, voluntary professional relationships with him should not be deemed unethical."

The AMA no doubt looked to the possibility that other state osteopathic societies might follow California's lead. In response, the AOA House of Delegates, under the leadership of President Charles L. Naylor, D.O., of Ravenna, affirmed its intention to maintain the status of osteopathy as a separate and complete school of medicine, vowing to "resist all efforts to be absorbed, amalgamated or destroyed."[14] When the Ohio State Medical Association (OSMA) issued its own policy statement, advising members to base their judgments "on knowledge of the professional, ethical and scientific competence of the individual doctor of osteopathy," Bill Konold distributed copies to OOA members. In a cover memorandum, he noted that the OSMA statement put in writing the "theoretical policy" under which M.D.'s and D.O.'s in certain parts of Ohio had been functioning

As president of the American Osteopathic Association in 1960–61, Charles L. Naylor, D.O., of Ravenna, affirmed osteopathy's intention to remain a separate school of medicine at a time when the AMA was pushing hard for amalgamation. *Photo by Fabian Bachrach, courtesy of Ohio Osteopathic Association.*

since 1943. "Our advice to you now," he wrote, "is to welcome quietly and personally the clarification. . . . The less public talk about it, the better. It's smart now, in our opinion, to let nature develop relations and progress on the proper basis." Konold promised to send each county medical society "pertinent facts about osteopathic medicine in Ohio and suggest a program for mutual understanding and living together for the benefit of community health and the advancement of patient care."[15]

By the mid-1960s, it had become clear that organized osteopathy was dug in and had no intention of amalgamating; no other state had followed California's lead. At its convention in Atlantic City in May 1967, the AMA baldly declared its intention to absorb the osteopathic profession by authorizing its board of trustees to begin negotiations with all osteopathic colleges for the purpose of converting them to orthodox medi-

Convention attendance set a new record, with more than seven hundred present, in 1958 when the Ohio Osteopathic Association marked its sixtieth anniversary. Among those attending were (*left to right*) Stephen J. Thiel, D.O., founder and medical director of Epp Memorial Hospital in Cincinnati, and former OOA presidents John W. Mulford, D.O., and Gertrud Helmecke Reimer, D.O. *Courtesy of Ohio Osteopathic Association.*

Among the leaders who helped the Ohio profession weather the threat posed by the California merger of 1961 were (*left to right*) John W. Hayes, D.O., Charles L. Naylor, D.O., Jack M. Wright, D.O., Donald C. Siehl, D.O., and Jack D. Hutchison, D.O. All five served as president of the Ohio Osteopathic Association. *Courtesy of Ohio Osteopathic Association.*

cal schools. The move only increased the resolve of the AOA leadership to remain separate. In a statement released from his home in East Liverpool, Ohio, and reported by the Associated Press, AOA President John W. Hayes, D.O., blasted the AMA. Its concern with the so-called "osteopathic problem," he said, was "a political problem . . . based on the fact that 13,000 osteopathic physicians, not under AMA control, are providing quality health care to some 20 million Americans, attaining growing public recognition and increased federal support."[16] An unsigned editorial in the *Buckeye* declared, "Once again, members of the profession can be grateful to the AMA; through their policy of negation they have stimulated and publicized the forward-looking and realistic view of the osteopathic profession toward the health care of the people."[17]

Ironically, in the midst of this interprofessional jousting, osteopathic physicians were quietly winning the respect of their M.D. colleagues. An important factor was the war in Vietnam. In 1956, near the close of the Korean conflict, President Eisenhower had signed Public Law 763, allowing graduates in osteopathic medicine to be appointed as commissioned medical officers in the armed services. In 1966, Harry J. Walter, D.O., of Bowling Green, volunteered for the U.S. Air Force, becoming the first osteopathic physician from Ohio to be commissioned under the new law. As the Vietnam War escalated and manpower needs increased, doctors—including, for the first time, D.O.'s—were drafted to serve. By 1971, over two hundred D.O.'s had served their country as medical officers of the armed forces, and another seventy-seven (eleven from Ohio) were due for induction that year.[18]

Richard L. Sims would later describe the salutary effect of war on interprofessional relations, observing that, as D.O.'s and M.D.'s served side by side in the military, "they developed a mutual respect which went a long way to break down old prejudices. . . . Young M.D.'s discovered that D.O.'s were very capable physicians and this word soon spread back to allopathic

James F. Sosnowski, D.O., a resident in internal medicine at Doctors Hospital, Columbus, was drafted to serve as a medical officer in 1967. He was killed the following year when a mortar shell hit his field hospital near Tay Ninh, South Vietnam. *Courtesy of Ohio Osteopathic Association.*

colleges and their teaching hospitals."[19] This new respect, however, soon challenged the unity of the osteopathic profession as allopathic hospitals began recruiting D.O.'s to their postdoctoral programs. In 1971, AMA President Richard L. Fulton, M.D., told the Ohio State Medical Association that many hospitals were unable to fill their quotas of allopathic residents and interns at the same time that many osteopathic trainees had no place to obtain additional training. "It seems only logical to me," he said, "that these same qualified physicians could train in our hospitals, become better physicians, be of value to the hospitals during their training years and end up being better practitioners. This would give our communities more well trained physicians to help in the delivery of medical care—which is the goal of all of us."[20] In 1967, restrictions banning osteopathic hospitals from membership in the Ohio Hospital Association and the American Hospital Association were lifted. A year later, the AMA opened its membership to D.O.'s, and in 1970 the Ohio State Medical Association followed suit.[21]

Although prejudice had largely waned, D.O.'s still grappled with the issue of their own identity. With the push for higher educational standards, and with full and unlimited practice rights secured in most U.S. states, D.O.'s increasingly duplicated the role and services of M.D.'s. Indeed, osteopathy since the days of Andrew Taylor Still had undergone a huge evolutionary change. The colleges had fully integrated pharmacology and surgery into their curriculums and had reduced their emphasis on distinctively osteopathic methods; by the end of the 1950s, few articles in the *Journal of the American Osteopathic Association* even mentioned osteopathic manipulative therapy (OMT). Although the OOA continued to provide annual refresher courses on manipulative therapeutics,[22] many D.O.'s had drifted away from traditional osteopathic technique, to the dismay of the profession's elders. Osteopathic pioneer Dr. Hugh Gravett, still actively practicing at age eighty-seven, sent a letter to be read at the past presidents' meeting at the 1949 OOA convention. "Our real trouble," he wrote, "is, and always has been, [that] we have simply grown faster than we could be

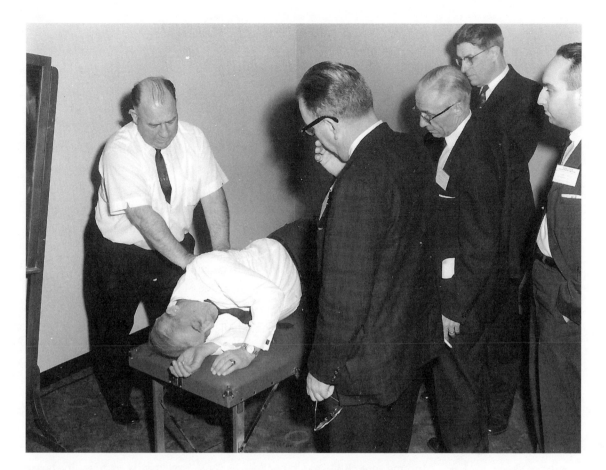

properly osteopathically managed. . . . Unless we Teach, Preach, and Practice that upon which osteopathy is founded, the wisdom of nature and our bodies, . . . neither we nor osteopathy will long survive."[23]

Questions of identity would persist as D.O.'s moved closer to M.D.'s in patient management. "Have we reached the crossroads?" OOA President Eugene R. DeLucia, D.O., a general practitioner from Warren, provocatively asked his colleagues in 1970. "How will we turn? What has happened to osteopathic medicine? Is the concept obsolescent? Has osteopathy been replaced by antibiotics and modern drug therapy? Is osteopathy synonymous with physical therapy? Has the philosophy of structural integrity and health become passé? Too laborious? Have our physicians found an easier way to practice?"[24] In his final message as OOA president, DeLucia urged a "renaissance" of osteopathic medicine, theory, and practice. "We have so well succeeded in proving the equality of osteopathic medicine that we have created a new problem. We must now prove that we are still a little different."[25] AOA President Donald Siehl, D.O., of Dayton, revived the theme a decade later. Addressing the national convention in 1979, he decried the

Physicians participate in the annual "fall refresher" on manipulative therapeutics, about 1955. *Courtesy of Ohio Osteopathic Association.*

number of D.O.'s who viewed the clinical application of osteopathic structural diagnosis and manipulative therapy as "a specialty function only." Siehl urged his colleagues to use the principles of osteopathy and become more proficient in these areas of their practice. "There are new things in manipulation," he said. "There are new things in structural diagnosis. There are new ways of treating the patient [using] the holistic approach."[26]

American medicine, meanwhile, was in flux. New federal legislation guaranteeing health insurance for the elderly and the poor was revolutionizing the delivery of health care and the financial arrangements for hospital care. Legislation enacting Medicare and Medicaid, passed in 1965 and implemented in 1966 as part of President Lyndon B. Johnson's Great Society program, represented the culmination of a twenty-year legislative debate over a program first proposed by President Harry S. Truman.[27] Covering most persons aged sixty-five or older, the Medicare program consisted of two related health insurance plans: a hospital insurance plan and a supplementary medical insurance plan. Medicaid was a health insurance program established for low-income persons under age sixty-five and persons over that age who had exhausted their Medicare benefit. Both programs would quickly grow larger than originally expected and lead the federal government to institute cost-containment measures that would dramatically affect doctors and hospitals.

In June 1968, Bill Konold resigned as OOA executive secretary. The board of trustees voted to retain his services as a consultant. William S. Konold & Associates would continue to publish the *Buckeye*, manage OOA conventions, and perform other public relations functions for several more years. To succeed Konold, the board selected Richard L. Sims, Konold's longtime assistant and a known quantity. Sims, a graduate of Ohio State University with a degree in business administration, had worked for Konold and the OOA since 1951, with time out for military service during the Korean War. Like his boss and mentor, Sims also wore several hats, serving as assistant administrator of Doctors Hospital and executive secretary of the Ohio Osteopathic Hospital Association at the same time he served as assistant executive secretary of the OOA. That arrangement continued: Dick Sims went on consulting while he monitored the pulse of the profession from Bill Konold's legendary red chair—"the hub," Sims called it, owing to his predecessor's twenty-nine years' occupancy.[28]

One of Sims's first projects as executive secretary was an image-improvement study. Public perception—or, more accurately, mispercep-

OOA Executive Secretary William S. Konold and James O. Watson, D.O., share a light moment at the 1968 convention, Konold's last as OOA executive secretary. *Courtesy of Ohio Osteopathic Association.*

In 1968, Richard L. Sims (*left*) succeeded William Konold as OOA executive secretary. Like his predecessor, Sims wore several hats simultaneously: OOA executive secretary, administrator of Doctors Hospital, and executive secretary of the Ohio Osteopathic Hospital Association. With Sims are J. David Luckhaupt, assistant administrator of Doctors Hospital, and Harold E. Clybourne, D.O., one of the hospital's founders. *Courtesy of Ohio Osteopathic Association.*

tion—of osteopathy had frustrated the profession from its very beginning. The small numbers of D.O.'s, compared to M.D.'s, and their disproportionate distribution (in some places osteopathic medical care was simply unavailable) helped make the profession "socially invisible."[29] Longstanding discrimination had also eroded the profession's standing with the public. Finally, the task of educating patients about this alternative approach to healing could be vexing, to say the least. "During my trips around the state," OOA President Leonard D. Sells, D.O., confessed in 1960, "I have been made aware of a recurring public relations problem . . . how to explain osteopathic medicine in simple terms to the layman. Our profession is still something of a mystery to segments of the general public, who wants to know, 'What is the difference between a D.O. and an M.D.?'" The D.O., said Sells, was still sometimes confused with the chiropractor, the chiropodist, and the optometrist.[30]

In 1968, the Ohio Osteopathic Association engaged two Ohio State University marketing professors, Roger D. Blackwell and James F. Engel, to conduct a study of how individuals select and retain a family doctor, and to measure the image of D.O.'s

Roger D. Blackwell, professor of marketing at Ohio State University, led a 1968 survey of the public image of D.O.'s versus M.D.'s. He found that most respondents were not negative toward D.O.'s, they were simply uninformed about them—a perennial problem for the profession. *Courtesy of Ohio Osteopathic Association.*

versus M.D.'s. With the help of sixty-two graduate students in the OSU College of Administrative Sciences, they selected a sample of 619 households in central Ohio—468 from Columbus and 151 from Dayton. The respondents were interviewed by telephone. The students then compiled and analyzed the results of the forty-seven-question survey. The resulting two-hundred-page report—the *Buckeye* called it "the finest set of factual statistics ever compiled about the image of osteopathy"— found that most respondents were not negative toward D.O.'s, they were simply uninformed about them. Only 20 percent of respondents could ac- curately define "osteopathy," while less than a third were aware of the equivalency of the M.D. and D.O. in training, state examinations, hospital facilities, and specialties offered. "What most of us have known for years," Sims wrote in the *Buckeye*, "is now substantiated beyond any doubt." The poll would provide guidance in mapping a program for the future; three decades' battle for equal recognition for osteopathy now would give way to a new campaign. "One thing is certain," he wrote, "the image of osteopa- thy needs improvement."[31]

Following the OSU image study, the Ohio Osteopathic Association embarked on a $15,000 image-enhancement campaign, hiring Paul and John Kaltenbach, an Akron advertising agency, to develop a multifaceted program. In the summer of 1970, the OOA unveiled a new exhibit at the Ohio State Fair, where it showed a film about the profession and distributed a new booklet, "Tools Are Made, Hands Are Born." The organization adopted a new logo and sent thirty-second public service announcements to each district academy for distribution to local television stations. "In the immediate future," the *Buckeye* reported, "the results of the OOA Image Program will be difficult to measure, but as we all know, Ohio has taken a very positive and necessary step in closing the knowledge gap . . . between our profession and the public."[32]

As the image program moved forward, OOA Public Relations Direc- tor Rick Weisheimer, together with writers Mary Jane Carroll and Chip

Elliott and designer Chris Snell, gave a fresh look to the *Buckeye* and produced a series of booklets designed to recruit new osteopathic graduates to Ohio. Dick Sims, meanwhile, focused on opening doors for D.O.'s with an aggressive campaign to gain the recognition of the state's health and public welfare departments. Drs. Arnold Allenius and Robert Turton became the first D.O.'s to be appointed (by Democratic and Republican governors, respectively) to the Ohio Public Health Council. Turton would go on to serve as chairman of the council, an advisory body to the Ohio Director of Health. With old barriers broken, osteopathic representation on matters of state health policy became the norm. The increasing respect the profession had earned during the postwar period would serve it well as Ohio D.O.'s pressed their next campaign: a state-supported college of osteopathic medicine.

5

AN OHIO COLLEGE OF OSTEOPATHIC MEDICINE

Be it resolved that the OOA form a committee to examine the
feasibility of establishing an opportunity for students to
pursue the D.O. degree within the state of Ohio.
—*OOA House of Delegates resolution, June 11, 1972*

IN JUNE 1973, AS THE OHIO Osteopathic Association marked its diamond anniversary at the Sheraton Columbus Motor Hotel, Dick Sims announced that he would resign as OOA executive director to become the full-time administrator of Doctors Hospital. Both Sims and the OOA board agreed that the professional association had grown and now required the services of a full-time executive director.[1] Christian H. Kindsvatter, a native of Wooster, Ohio, who had formerly held executive positions with the Ohio Retail Merchants Association and the Ohio Home Builders Association, took the reins on January 1, 1974. Consulting arrangements with Bill Konold and Charles Chastang came to an end as Kindsvatter assembled his own team. Helen Palmer became full-time OOA board secretary and convention manager. George F. Dunigan Jr., a government and geography teacher for the Logan schools and member of the Logan City Council, joined the staff as director of governmental affairs. And in April 1975, Jon F. Wills was hired as director of public relations. Wills, a graduate of Ohio University who had previously served as chief writer and publications coordinator for the Ohio Department of Transportation, would be responsible for press relations and internal communications, and serve as editor of the *Buckeye*.

At the top of the OOA agenda was the campaign for an Ohio college of osteopathic medicine. The osteopathic colleges had always been viewed as critically important in sustaining the growth and development of the profession. But for much of its history, the Ohio profession had been hobbled by the fact that young Ohioans desiring to become osteopathic physicians had to leave the state to obtain their education; all too often, they did not return. Speaking to Ohio students at the Chicago College of Osteopathy in 1938, M. A. Prudden, D.O., had complained, "We have for years sent more students to the osteopathic colleges than any other state that does not have an osteopathic college within its boundaries, and yet in spite of this, the number [of new doctors] coming into the state has gradually fallen. . . . If osteopathy in Ohio is to keep up, this condition must be changed."[2] By 1959, Ohio had become one of the top states providing students for the nation's six osteopathic colleges; of nearly two thousand students enrolled nationally, Ohio, with 148, ranked fifth.[3] Other factors, too, underscored the need for a new college: the loss of the College of Osteopathic Physicians and Surgeons in Los Angeles (which had become an allopathic medical college), a growing shortage of general practice physicians, and a declining ratio of D.O.'s to the U.S. population.[4]

In 1963, the Michigan Association of Osteopathic Physicians and Surgeons announced plans for the development of a new state-supported college of osteopathic medicine. Their Ohio counterparts watched the project with keen interest, even forming an Ohio support committee, chaired by Robert J. Kromer, D.O., of Sandusky, to help raise money for the new college.[5] In 1965, the Michigan Legislature provided money for a feasibility study. Despite organized opposition from the Michigan State Medical Society, a bill to establish the Michigan State University College of Osteopathic Medicine (MSUCOM) was finally approved in July 1969, and in September the new college admitted its first class of twenty.[6] Affiliated with Michigan State University in East Lansing, MSUCOM was the first new college of osteopathic medicine to be established in several decades. More importantly, it was the first university-based osteopathic college and the first state-supported osteopathic college. It thus represented a giant step forward for the profession and became a prototype for the subsequent establishment of other schools in Texas (1970), Oklahoma (1972), and West Virginia (1972). The establishment of MSUCOM also sparked discussion among Ohio D.O.'s: if Michigan could start a college, why not Ohio?

At the 1972 convention, at the urging of Donald C. Atkins, D.O., a Sandusky pathologist, the OOA House of Delegates approved a resolution authorizing a feasibility study for an osteopathic college in Ohio. Chaired by Martin E. Levitt, D.O., of Dayton, the college development committee plunged in, contacting state officials and legislators and making a presentation to Governor John J. Gilligan's Task Force on Health Care. On the committee's recommendation, the OOA hired education consultant Floyde E. Brooker to prepare a preliminary statement of the case for an osteopathic medical college in Ohio. In June 1973, the OOA House of Delegates voted a one-year, $50 assessment of each member to fund conclusion of the feasibility study and begin development of a new college. The funds would be deposited in the Ohio Osteopathic Foundation, established in 1963 but largely inactive prior to the serious drive for a new college.

College efforts suffered a setback when the Governor's Task Force on Health Care failed to ratify an OOA-sponsored amendment. It read, "The Task Force recommends that the Ohio Board of Regents be encouraged to consider the establishment of an osteopathic college of family practitioners in consortium with and to share facilities with a state-supported medical school." Nevertheless, the OOA pressed on. "It is becoming increasingly apparent," the *Buckeye* declared, " . . . that a college of osteopathic medicine in Ohio is not only a good idea, but is a necessity if Ohio is to curb a severe shortage of physicians and provide the health care the people of Ohio need and deserve."[7]

Under the leadership of Chris Kindsvatter and George Dunigan, the OOA began a multi-pronged effort to establish a state-supported college. To line up legislative support, the OOA commissioned a college feasibility study, prepared by William S. Konold and Associates, and sponsored an educational visit to Michigan State University that included a tour of MSU-COM led by Myron Magen, D.O., dean of the college. Ohio Senators Harry Meshel and Paul Gilmore, and Representatives Michael Del Bane, Tom Fries, and Corwin Nixon attended, along with Kindsvatter, Dunigan, and OOA President-elect Martin Levitt. After seeing how MSU had renovated unused dormitory space to house the osteopathic college, the legislative team became convinced that a similar model might work in Ohio.

The original feasibility study, which called for a new osteopathic college to share facilities with an established Ohio medical college, was revised to urge instead the development of a new college of osteopathic medicine at Ohio University in Athens. A college in southeast Ohio, the revised study said, could help alleviate a critical doctor shortage in that part of the state and at the same time attract an influx of federal funds to the economically depressed area from the Appalachian Regional Commission. The report also emphasized that Ohio University, which was then experiencing a severe decline in enrollment, already had space available for classrooms in vacant dormitory facilities—thereby precluding the need for costly new construction—and that the existing premedical staff could serve as a foundation for the college. Finally, southeast Ohio was the only area of the state not already served by a medical school. A college of osteopathic medicine might fill this void and, at the same time, help attract general practitioners to an underserved area.[8]

OOA Executive Director Christian H. Kindsvatter (*standing, left*) confers with Reps. Robert A. Nader (*seated*), Tom Fries, and George Tablack on House Bill 229 calling for creation of an Ohio college of osteopathic medicine, 1975. *Courtesy of Ohio Osteopathic Association.*

On January 30, 1975, Rep. Thomas Fries, Democrat of Dayton, introduced House Bill 229, which called for creation of an Ohio college of osteopathic medicine at Ohio University, in the 111th General Assembly. Fries served as primary sponsor in the House. Working with the OOA as primary sponsor in the Senate was Neil F. Zimmers, Democrat of Miami and Montgomery Counties. To prod the Ohio Legislature to approve establishment of the new school, the OOA House of Delegates took a bold next step, voting to assess each of its 1,021 members $250 annually for six years "for the purpose of establishing and perpetuating an Ohio college of osteopathic medicine."[9] OOA Executive Director Chris Kindsvatter called the move "tremendous and unprecedented. . . . We're the first professional association in the state to ever make this kind of an offer."[10]

Opposition to the proposed school came principally from Ohio Board of Regents Chancellor James A. Norton, who said he did not see a place for such a school in the list of the state's priorities for higher education. The *Cleveland Plain Dealer* agreed. "Enough is being spent to assure family doctors through new curriculums in the existing medical schools," the newspaper argued, adding that a new school would be beyond the state's means.[11] But the *Athens Messenger* took another view. It declared, "Young Ohioans who desire to become osteopaths must now leave the state for their education; too often, they do not return, depriving all Ohio residents of the general practitioners' services that are already in short supply throughout the state and are most particularly scarce in southeastern Ohio." The editorial noted that in the seventeen counties of southeast Ohio there were just

512 doctors to serve 700,000 persons. Osteopathic physicians, the *Messenger* argued, "provide the 'family doctor' kind of medicine that is especially lacking in southeastern Ohio; the entire state could use more general practitioners to offset a long trend toward medical specialization."[12]

The pending legislation, according to OOA President Martin Levitt, had brought Ohio D.O.'s to yet another decisive point in their professional evolution. "This time," he told the 1975 convention, "we have been set upon by our opposition not because of our differences but because of our similarities. Those who would defeat us have stated that the existing schools are more than adequate. . . . Today, then, [an Ohio college of osteopathic medicine] has come to be a symbol of our rights to exist, perpetuate our profession and provide for that segment of the population served by our physicians." He concluded, "Let us utilize this opposition as we have in years prior to unify and motivate us."[13]

As it turned out, the legislation enjoyed smooth sailing. In March, the Ohio House of Representatives passed the bill by a vote of 90 to 4. The Senate—where it had the unqualified support of Finance Chairman Harry Meshel, Democrat of Youngstown—added its approval, 24 to 6, in July. And, on August 18, 1975, despite intense pressure from the Ohio State Medical Association to veto it, Gov. James A. Rhodes signed the bill creating the Ohio University College of Osteopathic Medicine (OU-COM), Ohio's seventh college of medicine. Dr. James Watson, the architect of the Ohio profession's preeminent legal victory giving D.O.'s full and unlimited practice rights, would later call it "perhaps the most important and significant legislative event since the revision of the medical practice act in 1943."[14]

Ohio University President Harry B. Crewson had been an early supporter of the college. Together with Martin L. Hecht, vice president for university relations, he had worked closely with OOA leadership, quietly registering the university's support as House Bill 229 made its way through the legislature. "I was for the idea immediately," Crewson later recalled. "It would obviously help the university and the community greatly, and add a new dimension to our faculty. With the enrollment decline, there was also a need for anything that would boost campus morale."[15] In a memo to Dr. Charles J. Ping, shortly due to succeed him as O.U. president, Crewson urged Ping to "begin immediately to develop plans and procedures for implementation" of the college.[16]

The legislation authorizing the new college called for creation of a ten-member advisory board and appropriated $670,000 to plan for the opening of the new school, which looked toward admitting its first class for the 1976–77 academic year. Gerald A. Faverman was named acting dean;

Between mid-November 1975 and September 1976, Acting Dean Gerald A. Faverman worked tirelessly to ready the new college to receive its first students. *Courtesy of Ohio University College of Osteopathic Medicine.*

Faverman was assistant dean of MSUCOM and had worked as a consultant to OOA on the college feasibility study. Frank W. Myers, D.O., director of the general practice residency program at Brentwood Hospital and a general practitioner in Northfield, was named associate dean for clinical affairs with responsibility for developing and supervising clinical education.

Faverman arrived at OU-COM in November 1975. Ohio University President Charles J. Ping, like Crewson a staunch supporter of the new college, would later recall that Faverman "took charge like a whirlwind."[17] By January 1976, tuition had been set; curriculum, admissions, and scholarship advisory committees had been created; twelve faculty members had been appointed, and space planning was under way. Working closely with Faverman to launch the new college was Charles A. "Chip" Rogers Jr., assistant to the dean. Rogers would have a long and distinguished career at OU-COM, later serving as director of alumni relations and director of advocacy—an indispensable "ambassador" whose service would be recognized with the OOA Meritorious Service Award in 1994.

In an extended interview published in the *Buckeye,* Faverman shared his plans for the college. Asked if he intended to pattern OU-COM after Michigan State's College of Osteopathic Medicine, Faverman told the *Buckeye* that he intended to borrow the best innovations from every school—the laboratory design at Oklahoma, the rural clinics at Kirksville—to create something uniquely Ohio's. Asked what kind of student the school sought to attract, Faverman said, "We intend to discriminate in favor of Ohio residents. Eighty percent of the class is allocated to Ohio residents or to others from out of state who sign a statement of intent to practice in Ohio upon

OU-COM's first class of twenty-four students arrived in September 1976. *Back row, left to right:* Gary Welch, Jules Sumkin, Mark Thompson, Ron Moomaw, Mac Poll, Laurie Serif, John Ragazzo, Paul Haupt, Margaret Fankhauser, Robert Biscup, Randy Larrick. *Front row, left to right:* John Reed, Stephen Evans, Will Schlotterer, Mark Pace, Stephen DuBos, Audrey Johnson, Stephanie Knapp, Hinda Abramoff, Patricia Dempsey, Wendy Ngo, Joan Resatka, Suzanne Kimball, and Ted Ivanchak. *Photo by Jon F. Wills, courtesy of Ohio Osteopathic Association.*

graduation." The school, he said, would pay "strong attention" to those who planned to go into primary care in a rural or Appalachian environment. Finally, he said, the college would "pay heed . . . to those who have some true understanding of osteopathic medicine and wish to be osteopathic physicians as a first choice."[18]

Construction crews were still hammering in Grosvenor Hall, a former dormitory hastily remodeled for classroom use, when the first class of twenty-four students arrived in September 1976: sixteen men and eight women representing thirteen Ohio counties, as well as the states of Michigan and New York, selected from among 284 applicants. The first two years of the four-year curriculum would be centered on the Athens campus; introduction to clinical training would be carried out in existing facilities in southeast Ohio. Third- and fourth-year students, meanwhile, would receive varied clinical training in general and community osteopathic hospitals, rural clinics, and physicians' offices throughout the state, rotating within

Ohio University President Charles J. Ping presides over the dedication of the new Ohio University College of Osteopathic Medicine, September 11, 1976. Behind him is Grosvenor Hall, the first college building. The former dormitory was still being re-modeled for classroom use when the first class of twenty-four students arrived that year. *Photo by Jon F. Wills, courtesy of Ohio Osteopathic Association.*

one of seven regional teaching centers to provide maximum exposure to both specialty and family practices in the affiliated hospitals. The brainchild of Acting Dean Gerald Faverman, the regional teaching centers provided an innovative framework for osteopathic training, offering a range of clinical learning environments, from large tertiary care medical centers to small community hospitals. Each region would have its own dean and its own director of medical education but the same pioneering "continuum" curriculum, blending the basic sciences, clinical training, medical ethics and humanities, and osteopathic principles and practice. The regional structure

Gerald A. Faverman (*right*), acting dean of OU-COM, unveils a plaque honoring the contributions of the osteopathic profession to the support of the college during convocation ceremonies, 1976. With him are OOA Executive Director Christian H. Kindsvatter (*left*) and Martin E. Levitt, D.O., who chaired the OOA College Feasibility and Development Committee. *Courtesy of Ohio University College of Osteopathic Medicine.*

envisioned and implemented by Faverman would serve as an important foundation for future development of the college, eventually leading to formation of the college's Centers for Osteopathic Research and Education (CORE), a statewide consortium of postdoctoral training programs.

By March 1976, Selby General Hospital in Marietta and Grandview Hospital in Dayton had signed affiliation agreements with OU-COM to develop and operate teaching programs for students of the new college. By June, the college had affiliated with nine hospitals having a combined total of more than 2,200 beds. In addition to Selby and Grandview, they were Veterans Memorial, in Pomeroy; Northeastern Ohio General, Madison; Athens Health and Mental Retardation Center, Athens; O'Bleness Memorial, Athens; Brentwood, Cleveland; Youngstown Osteopathic Hospital; and Doctors, Columbus.

OU-COM signs an affiliation agreement with Dayton's Grandview Hospital, 1976. *Left to right:* Acting Dean Gerald A. Faverman; John R. Shryock, president of the Grandview board of trustees; O.U. President Charles J. Ping; and Richard J. Minor, Grandview president and chief executive officer. *Courtesy of Ohio University College of Osteopathic Medicine.*

In December 1976, OU-COM appointed a committee to begin planning for the school's second medical building. Irvine Hall, a dormitory, would be renovated to house basic life science instruction, biomedical research, and clinical demonstration. The new building would permit a major increase in enrollment. In September 1977, the college admitted its second class, comprising thirty-six students, and in November, President Ping

Lecture in Irvine Hall, the second building renovated for the medical college. *Photo by Sandy Lungershausen, courtesy of Ohio University College of Osteopathic Medicine.*

Associate Dean Frank W. Myers, D.O. (*right*), was named dean of OU-COM in November 1977. Standing with him during the 1980 OOA convention is Ralph S. Licklider, D.O., one of the profession's patriarchs. *Courtesy of Ohio University College of Osteopathic Medicine.*

named Associate Dean Frank Myers as dean of OU-COM. In September 1978, the college admitted forty-eight new students. Enrollment projections called for four hundred students within ten years.

In the spring of 1980, the Ohio University College of Osteopathic Medicine graduated its first class of twenty-one seniors and prepared to welcome a new entering class of seventy-two. There were now thirty-one full-time physicians on the clinical faculty. A second building, Irvine Hall, was slated to open in October, providing new basic science offices and laboratories and two lecture halls, and specifications were in hand for a new ambulatory care center in Parks Hall funded jointly by the state and the Appalachian Regional Commission. Completing an important piece of the puzzle, Doctors Hospital of Columbus took possession of the eighty-five-bed Mount St. Mary Hospital in Nelsonville, twelve miles away. Renamed Doctors Hospital of Nelsonville, it would play a key role in the clinical training of OU-COM students. By the end of 1980, the OOA, through the annual assessment of its

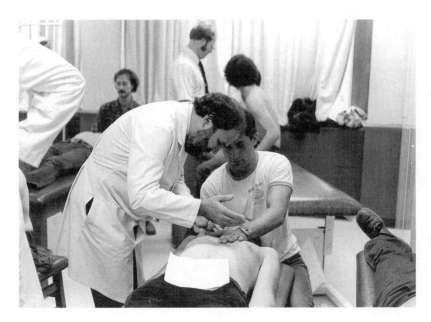

Anthony G. Chila, D.O., associate professor of family medicine at OU-COM, instructs students in OMT technique, date unknown. *Courtesy of Ohio University College of Osteopathic Medicine.*

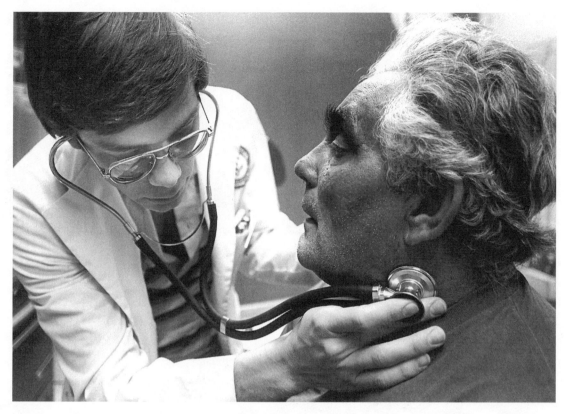

Third-year student Ed Tappel examines Walter Tabler, a member of the Kilvert Indian community, 1985. *Courtesy of Ohio University College of Osteopathic Medicine.*

members, had funneled more than $850,000 to the fledgling college. The money had been used for critical equipment needs and to increase faculty slots. Having earned full accreditation from the American Osteopathic Association, OU-COM was authorized to increase its September 1982 entering class to one hundred, representing full enrollment.

By early 1983, a new osteopathic medical center in Parks Hall, staffed by OU-COM clinical faculty and physician administrators, was treating more than one hundred patients from southeast Ohio each day.[19] Word of the Ohio osteopathic college spread with

Since 1980, the daily radio program *Family Health*, produced at OU-COM and widely syndicated, has provided health information and spread news of the college. Frank W. Myers, D.O., dean of OU-COM, was program host from 1983 until 1998. *Courtesy of Ohio University College of Osteopathic Medicine.*

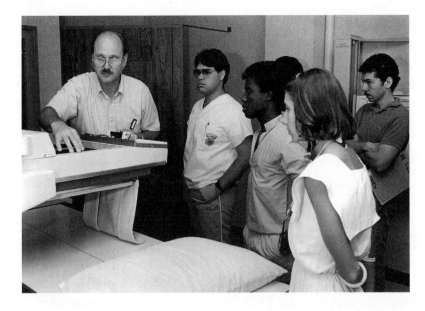

Frederick Rente, D.O., professor of radiology, explains the use of imaging equipment to participants in Summer Scholars, 1985. The program was begun to boost the number of minority students enrolled in the college. *Courtesy of Ohio University College of Osteopathic Medicine.*

the broadcast of *Family Health*, a daily two-and-one-half-minute radio program on health care produced by OU-COM with the Ohio University Telecommunications Center. Initially hosted by Fred Jensen, D.O., and later by Dean Frank Myers, the series began in 1980 with eighteen radio stations in southeast Ohio and West Virginia; by 1989, 460 stations in 49 states carried the program.[20] With yet another initiative, the Summer Scholars program, the college hoped to address the trouble osteopathic

Sandra Biechler, D.O., of Massillon, assists medical student Ashley Williams at the White Coat Ceremony, 1998. The Arnold P. Gold Foundation White Coat Ceremony was initiated in 1993 at Columbia University and soon after at many other medical schools, including OU-COM. This important "rite of passage," which occurs as medical students begin their clinical training, is meant to encourage scientific excellence and compassionate care in the practice of medicine. *Courtesy of Ohio University College of Osteopathic Medicine.*

medical schools had long had in attracting qualified minority students and to change the face of what, until now, had been a largely white student body. Beginning in 1982, ten minority students—African American, Hispanic, and Native American—were selected to participate in an eight-week summer program designed to explore careers in osteopathic medicine and enhance their preparation for admission to osteopathic medical school.[21]

By 1991, 10 percent of D.O.'s practicing in Ohio were OU-COM alumni.[22] In December that year, Dean Frank Myers reported that 1,800 applications—the highest number on record—had been received for the class of 100 that had entered in September. The burgeoning interest was part of a national trend: in 1993, the American Association of Colleges of Osteopathic Medicine reported that its fifteen member colleges together enrolled 2,035 freshmen for the 1992–93 academic year, breaking the two-thousand barrier for the first time in the profession's history.[23] As the national numbers climbed upward, so did OU-COM's: for the 1994–95 academic year, the college received more than 3,900 applications for 100 first-year places— more than twice the number received just three years earlier.

In 1993, after launching a novel project to provide free immunizations

to all children in Appalachian Ohio—a plan U.S. Senator John Glenn called "a model for the nation"[24]—Frank Myers stepped down as dean to return to teaching and family practice. He was succeeded by Barbara Ross-Lee, D.O., formerly associate dean for health policy and a professor of family medicine at MSUCOM. Ross-Lee, a former Robert Wood Johnson Health Policy Fellow who had once worked for U.S. Senator Bill Bradley, Democrat of New Jersey, garnered much media attention as the first African American woman to head a U.S. medical school. She also brought new visibility to the college and the profession with her focus on health policy research and education as a means of solving the nation's health care problems.

Barbara Ross-Lee, D.O., was named dean of OU-COM in 1993. Innovative curriculum and a strong focus on health policy research and education as a means of solving the nation's health care problems marked Ross-Lee's eight-year tenure. *Courtesy of Ohio University College of Osteopathic Medicine.*

Barbara Ross-Lee, D.O., presents the Dean's Award to OU-COM graduate Jacqueline Heiser, D.O., 1997. *Courtesy of Ohio University College of Osteopathic Medicine.*

Under Ross-Lee's leadership, OU-COM and the state's medical educators forged a partnership that fundamentally changed osteopathic medical education in Ohio. In 1994, the college's seven regional teaching centers were reduced to five—the inevitable consequence of the consolidation or closure of so many of the state's independent osteopathic hospitals during this period—and renamed the Centers for Osteopathic Research and Education (CORE).[25] With its reshaped curriculum and use of high-tech distance-learning technologies, CORE became a national model for the delivery of medical education.[26] Ross-Lee also launched a health policy fellowship program in cooperation with the AOA to prepare osteopathic physicians for leadership roles in government and the profession, and established a new department of social medicine. Chaired by historian and sociologist Norman Gevitz, author of *The D.O.'s: Osteopathic Medicine in America* (1982), the new department was seen as a means of incorporating such topics as osteopathic medical history, sociocultural health issues, and health policy that were often left out of the curriculum. All these initiatives won OU-COM recognition as a leader in innovative curriculum. "She gave us a vision and laid the foundation for that vision," Jon Wills would later say of Ross-Lee.[27] Following her resignation in 2000, Daniel J. Marazon, D.O., served as interim dean, and in October 2001, John A. "Jack" Brose, D.O., a faculty member in the college's Department of Family Medicine since 1982 and assistant dean of clinical research since 1994, was named to succeed Ross-Lee.

OU-COM fulfilled its promise. By 2001, the college had graduated 1,649 D.O.'s. Sixty-four percent were practicing in Ohio; 56 percent of

OU-COM graduates, 1999. OU-COM ranks among the nation's top medical schools for producing family doctors. *Courtesy of Ohio University College of Osteopathic Medicine.*

those practicing were primary care physicians; 22 percent were serving the health care needs of towns with populations under 50,000. Statistics showed, too, that OU-COM—and, thereby, the profession itself—had grown in diversity: of 105 students in the entering class of 2000, 34 percent were female, while 24 percent were minorities, including African Americans and Hispanics.[28] The 2001 American Medical Student Association Foundation's Primary Care Scorecard, used by premeds to determine which schools foster primary care ideals, ranked OU-COM second among the nation's 144 medical schools in the percentage of its 1999 graduates entering primary care residency programs (72 percent).[29] Former OOA Executive Director Richard Sims would later call the creation and development of OU-COM "one of the proudest chapters in the history of the Ohio Osteopathic Association."[30]

6

A SECOND VOICE

> Osteopathic hospitals are being asked why an *osteopathic* hospital?
> Why *osteopathic* specialist? Why *osteopathic* anything?
> —*Unsigned AOA editorial, reprinted in the*
> Buckeye Osteopathic Physician, *January 1974*

BY 1979, THERE WERE APPROXIMATELY fourteen hundred D.O.'s practicing in Ohio. Most were engaged in general or family practice, leading the Ohio Osteopathic Association to estimate that the state's osteopathic physicians provided a "disproportionately high portion" of all primary health care in Ohio. As they had historically, many practiced in rural areas; in some counties they were the only doctors. In 1979, there were fifteen osteopathic hospitals in Ohio, fourteen of which were accredited for intern and residency training. In addition, osteopathic physicians served on the medical staffs of 139 mixed-staff (M.D.–D.O.) hospitals.[1]

The establishment of OU-COM had brought a tremendous sense of pride and accomplishment to the state's professional association. But the campaign had not been without cost. Feasibility-study and lobbying expenses had eaten badly into the OOA's operating budget, and by the middle of 1975 the organization faced mounting financial problems. Its reserves were depleted and it faced a growing deficit. A central problem was that, of the $500 each member paid in annual dues, only $150 could be used for operating expenses; the rest was placed in liability accounts and distributed annually to the colleges, including to OU-COM. An austerity program, together with an increased allocation from dues for central office support (boosted to $250), helped put the organization back on a sound footing.

In 1977, Jon Wills, director of public relations, was named executive director. At the 1978 OOA convention, Wills told the delegates that one of the biggest problems facing the profession continued to be public relations. He noted that Ohio's osteopathic physicians "have accomplished many

things which have outstanding potential in the eyes of the public." As examples, he cited their growing involvement in the development of health maintenance organizations (HMOs), the establishment of a new ambulatory care center at Grandview Hospital—"widely recognized as an international prototype"—and a new emergency intermediary service in Lorain. "The profession is on the threshold of becoming a significant second voice in medicine," he concluded. "We must respond to this role. If we do not, we may just be wasting our time."[2]

Jon F. Wills was named executive director of OOA in 1977, when Ohio—and the nation—were on the cusp of unprecedented change in the delivery of health care. Here, he chats with Mary L. Theodoras, D.O., of Dayton, during "Elizabethan Fun Night" at the 1984 OOA convention in Cleveland. *Courtesy of Ohio Osteopathic Association.*

With Wills at the helm, the OOA regained financial stability and by 1981 the professional association again enjoyed a modest cash reserve. Despite its small staff, it continued to represent the profession's interests before myriad state health care agencies and commissions. In 1980, Wills reported, Ohio had more D.O.'s serving on government boards and commissions than at any other time in the profession's history. But increased government monitoring and control of health care delivery in an attempt to control costs was also complicating life for osteopathic medicine, especially in states like Ohio deemed to have too many hospital beds. New health planning laws now charged state and local agencies with developing crite-

OOA annual banquet, Cincinnati, 1977. Though buoyed by the successful launch of the Ohio College of Osteopathic Medicine, the profession would soon be at a turning point, facing an increasingly competitive health care environment. *Photo by Mayhew Photographers, courtesy of Ohio Osteopathic Association.*

The Ohio Osteopathic Medical Assistants Association

Rendering Honest, Loyal, and Efficient Service

In 1968, the OOA Board of Trustees voted to create a committee to organize and oversee a statewide osteopathic medical assistants association. Committee members included Arnold O. Allenius, D.O., of Columbus, Frank W. Myers, D.O., of Cleveland, and G. Joseph Strickler, executive director of the Dayton Academy of Osteopathic Medicine. The committee drafted a constitution and bylaws, and on September 20, 1969, the OOA board granted the Ohio Osteopathic Medical Assistants Association (OOMAA) an affiliate charter.

OOMAA was organized on November 9, 1969, in Dayton, where the medical assistants ratified a constitution and bylaws and elected Janet Bragg, an employee of the Nordonia Hills Clinic in Northfield, as the organization's first president. Other officers included Gretchen DeVoe, Columbus, president-elect; Linda Sauer, Dayton, vice president; Dorothy Jones, Cleveland, recording secretary; and Diane Hensley, treasurer. OOMAA's bylaws stipulated its purpose: to inspire its members to render honest, loyal, and efficient service to the osteopathic profession; to cooperate with the osteopathic profession in improving the public relations of the osteopathic profession; and to provide educational services to and on behalf of its members. In 1972, the fledgling organization hired John Robson, executive secretary of the Cleveland Academy of Osteopathic Medicine and the OOMAA Cleveland chapter, to serve as executive secretary. Robson held the position until 1982, when he handed the reins of OOMAA management to the Ohio Osteopathic Association.

OOMAA members played a notable role in the creation of the Ohio University College of Osteopathic Medicine, writing letters, making phone calls, and finding patient advocates to support the cause. In 1977, they voted to establish an annual scholarship to support a medical student at OU-COM, calling on each chapter to hold at least one fund-raising event each year, with proceeds earmarked for the scholarship. In 1983, the OOMAA board voted to establish a statewide "Boss of the Year" award. At its fifteenth annual convention, in 1984, OOMAA presented the first such award to Dr. Arnold Allenius, who had chaired the OOMAA Advisory Committee for thirteen years.

OOMAA membership peaked in 1983, with 374 members. As OOMAA grew, its membership became increasingly eclectic, including everyone from office receptionists to physician assistants, nurses, and X-ray technicians. It became increasingly difficult to plan programs to meet the needs of all of its members. The leadership tried for several years to establish a certification program for medical assistants, but was unable to obtain the necessary funding. As the founding office managers retired, newer employees gradually lost interest in OOMAA membership. Soon the only active chapter of the organization was the Southern Ohio Osteopathic Medical Assistants Association, composed largely of OU-COM employees. OOMAA became inactive in 1993, just shy of its twenty-fifth anniversary.

ria for the delivery of health care services, requiring hospitals to obtain a "certificate of need" from the Ohio Department of Health in order to expand, modernize, or purchase costly new equipment. Local health planning agencies, which were empowered to review the applications and make recommendations to the state health department, began turning down applications for new beds unless the need could be proven.

In 1970, the Metropolitan Health Planning Corporation (MHPC), an agency serving a five-county area of northeast Ohio, rejected Bay View

Hospital's application for expansion and renovation, and instead called for Bay View to close and to combine with one or more West Side hospitals to build a new joint-staff facility. In rejecting Bay View's application, the corporation said, "it is imperative that osteopathic and medical physicians practice within the same institution." To complaints from the Cleveland Academy of Osteopathic Medicine that the closure would harm osteopathic identity by "attempting to dictate where and with whom a group of physicians shall or shall not practice," MHPC Executive Director Lee Podolin did not mince words. "It's no concern of mine," he told the *Cleveland Plain Dealer*. "It's better than pouring eight or nine million bucks down a sewer. We can't continue a system based on prejudice or tradition." Osteopathic physicians clustering in their own hospitals, he said, "is no more useful than would a group practice be if it were limited to blue-eyed physicians."[3]

With certificate-of-need regulations threatening the very survival of the state's osteopathic hospitals, the American Osteopathic Hospital Association (AOHA) in 1978 called for an amendment to the national health planning law that would prevent discrimination against osteopathic hospitals and physicians. Without separate consideration, the AOHA maintained, discrimination would continue against hospitals serving D.O.'s and their patients.[4] In 1979, Congress amended the National Health Planning and Resources Development Act of 1974 to require local and state health systems agencies to review applications from osteopathic hospitals on the basis of the need for, and availability of, osteopathic services and facilities. State agencies would also be required to consider an application's impact on D.O. and M.D. training programs. The amendments signaled Congress's intent that osteopathic hospitals be considered equally with all others.

In 1978, only two years after Northeastern Ohio General Hospital completed a new federally funded wing that increased the number of beds from fifty to eighty-two, the MHPC recommended that the hospital merge with another Lake County hospital. Northeastern officials got help from the OOA and from its patients, who circulated petitions demanding the hospital's continued autonomy.[5] Many other osteopathic facilities, meanwhile, received a green light to expand their facilities and services. In 1980, open-heart surgery was approved for Doctors Hospital (Columbus) over the competing applications of two M.D. hospitals. In 1983, Brentwood Hospital in suburban Cleveland embarked on its fourth major expansion, and Marietta's Selby General Hospital broke ground for a $6.3 million expansion and renovation project. In 1984, Dayton's Grandview Hospital, Richmond Heights General Hospital, and Youngstown Osteopathic Hospital each

James O. Watson, D.O., and Evelyn L. Cover, D.O., 1983. Watson (1901–1985) was the first D.O. to serve on the State Medical Board of Ohio, Cover the third. Cover (1921–2001), who served from 1975 until 1983, was the first woman to serve on the board and the first D.O. to serve as board president. *Courtesy of Ohio Osteopathic Association.*

completed major expansion and renovation projects, and in 1987 Warren General Hospital completed a $10.4 million upgrade.[6]

In response to an increasingly competitive health care environment, the OOA once again turned its attention to the perennially vexing problem of public perception. In 1980, the first statewide public relations conference brought together PR representatives from the hospitals and from OU-COM. Now, the OOA laid the groundwork for an ambitious program to increase public awareness of the osteopathic profession. In a 1981–82 study funded by the OOA, Ohio University journalism professors Hugh M. Culbertson and Guido H. Stempel III compared the profession's perception of itself with that of the public. The study revealed that, while the D.O.'s surveyed were proud of osteopathy and saw it as an expanding profession, they also perceived a lack of public awareness of their field—a perception borne out by the public survey.[7] The sobering conclusion, Jon Wills reported, was that "only a massive advertising program over a period of time could achieve the expectations of the membership and educate the public about osteopathic medicine on a massive scale."[8]

With all but one of the state's fifteen osteopathic hospitals underwriting the project (Cuyahoga Falls elected not to participate), the OOA Advertising Committee selected Fahlgren and Ferriss of Cincinnati to plan an ambitious print, radio, and television campaign. The program would consist of a one-year test, using radio in the Dayton market and television in Toledo; pre- and post-awareness surveys would be conducted to assess the results. In addition,

The OOA advertising campaign of 1984–85 included this print ad, which ran in newspapers and magazines in selected markets. The campaign was in response to a 1984 survey that revealed a discouragingly low public awareness of osteopathy. *Courtesy of Ohio Osteopathic Association.*

newspaper advertisements would appear between December 1984 and April 1985 in all markets where a participating hospital was located (excluding Dayton and Toledo). The pre-awareness survey revealed a discouragingly low awareness of the title "Doctor of Osteopathy"—just 8 percent (versus 55 percent for "Doctor of Medicine")—and 56 percent for "D.O." (versus 99.3 percent for "M.D."). The post-awareness survey showed that radio was more effective than television. (The results of the print campaign were not measured.) But, while test marketing proved that advertising could help strengthen and improve public perception of D.O.'s, significantly increasing awareness among the general population would require a prohibitively expensive multiyear commitment.

The public relations dilemma was soon eclipsed by other problems, including a professional liability insurance crisis. Doctors were reeling from sharp increases in their medical malpractice insurance premiums, driven by large jury awards and settlements for victims of medical mistakes, as well as by the insurance companies' poor earnings on investments. Some insurance companies were getting out of the business altogether, leaving doctors to scramble to find coverage. In 1986, the OOA joined with the Ohio State Medical Association and the Ohio Hospital Association to form the Health Care Coalition for Professional Liability Reform to lobby for tort reform to limit awards for pain and suffering. Although the Ohio General Assembly eventually passed comprehensive tort-reform legislation, it was later declared unconstitutional by the Ohio Supreme Court, and insurance problems would continue to plague the state's doctors.

Meanwhile, a hospital economic crisis was affecting osteopathic and allopathic institutions alike. To rein in rising health care costs, the federal government in 1983 set standard payments for the care of patients with a particular diagnosis, paying doctors and hospitals according to a rigid, complex statutory formula that often bore little relation to the realities of local health care markets. In addition, alternative health care delivery systems, such as HMOs and preferred provider organizations (PPOs), combined with cutbacks in Medicare reimbursements and the change from an inpatient to outpatient environment, were causing economic distress for many small and mid-sized hospitals. "The very existence of many of our members," Richard A. Royer, president of the Ohio Osteopathic Hospital Association, reported in 1987, "is now threatened by massive economic forces that just a few years ago did not exist. Several of our members have merged, consolidated, or simply closed their doors to the public." Royer described the predicament hospitals were facing: "Third-party payers have sought to restrict access to hospitals by beneficiaries of their insurance plans. State officials [have] attempted to blur the distinct recognition gained by osteopathic hospitals. Managed care plans have also attempted to close our hospitals out of contracts for care of their patients. Most ominous of these, however, is the increasing trend of allopathic institutions competing not only for our patients, but our physicians as well." With an "alarming number" of D.O.'s no longer practicing in affiliation with osteopathic hospitals, Royer said, the situation presented a "call to action."[9]

In April 1988, declining occupancy caused Otto C. Epp Memorial Hospital in Cincinnati to merge with Midwest Health Systems, Inc. Later the same year, with its census down to seventeen patients a day, Northeastern Ohio General Hospital in Madison closed its doors, leading the OOA Board of Trustees to create a task force to monitor the survival of osteopathic hospitals in the state.[10] Meanwhile, facing double-digit increases in health insurance premiums, Ohio employers were pressuring insurers to lower health care costs. Ohio's Blue Cross and Blue Shield plans and health maintenance organizations turned to "selective contracting" as an answer to skyrocketing costs, demanding discounted fees from providers in return for exclusive contracts. Osteopathic hospitals and specialists soon found themselves at a distinct disadvantage: as the smallest institutions in the marketplace, and thereby lacking bargaining power, osteopathic hospitals were either excluded from negotiating with health plans or else offered less-than-favorable contracts. And osteopathic specialists, who by and large limited their practices to the osteopathic hospitals, were frozen out—general- and family-practice D.O.'s faced pressure to send their patients to larger M.D.

The closing of Northeastern Ohio General Hospital in Madison in November 1988 alarmed the state's osteopathic profession. *Courtesy of the* News-Herald, *Willoughby, Ohio.*

institutions with their predominantly M.D. specialists, instead of to osteopathic facilities.

To address this inequity, in early 1990 the OOA and the OOHA sought legislative relief. House Bill 593, sponsored by Rep. Wayne Jones, Democrat of Cuyahoga Falls, would require HMOs to pay for care their subscribers received at osteopathic hospitals, even if the HMO and the hospital did not have a contract, if the subscriber was admitted by a contracting physician. Payment to the noncontracting hospital would be mandated at the average charge for contracting hospitals in its service area, and the osteopathic hospital receiving payment would be required to accept it as payment in full. George Dunigan, OOA director of legislative affairs, actively lobbied for the bill, and the profession's leaders testified on its behalf before the House Health and Retirement Committee. "If HMOs, and other third-party payers in general, selectively exclude osteopathic hospitals from payment plans," Frank W. Myers, D.O., dean of OU-COM, pointed out, "they undermine the partnership which was established between the profession and the Legislature to insure the education of osteopathic physicians in this state." Mark C. Barabas, CEO of Youngstown Osteopathic Hospital, and Richard J. Minor, president of Grandview Hospital, testified that exclusion from HMOs was eroding the osteopathic hospitals' referral bases. Some HMOs, they charged, were contracting with osteopathic general practitioners but denying contracts to osteopathic specialists and hospitals, requiring that referrals be made to allopathic institutions instead.[11]

Although House Bill 593 passed the House by a vote of 94 to 4, it later died in the Senate Financial Institutions and Insurance Committee following intense opposition from the insurance industry. As a stopgap measure, the Ohio Senate passed an amendment creating a special study committee to review specific concerns of the osteopathic profession and make recommendations pertaining to the participation of osteopathic hospitals in HMOs. The eleven-member committee consisted of three Senate and three House members, the state superintendent of insurance, representatives of OOA and OOHA (Harold Thomas, D.O., of Cleveland, and Richard Minor, CEO of Grandview Hospital), and representatives of two managed-care organizations. Despite strong opposition from the HMO and insurance industries, the Ohio legislature voted to try a pilot project modeled on House Bill 593 at Grandview and Southview Hospitals, and to continue the Study Committee on Participation of Osteopathic Hospitals in HMOs. Although the OOA and OOHA were eventually able to secure passage of legislation prohibiting health plan discrimination against osteopathic hospitals and specialists, they were unable to persuade lawmakers to ensure that osteopathic health care systems were represented in all managed-care plans, and preferential contracts continued to bypass osteopathic hospitals and specialists.[12]

Despite the debilitating impact of managed care, the OOA and OOHA continued to work together throughout the 1990s to encourage an active partnership between D.O.'s and osteopathic hospitals. "D.O.'s are needed in osteopathic hospitals to help perpetuate the profession and keep the hospitals healthy," Youngstown Hospital CEO Mark Barabas wrote in a letter to the *Buckeye Osteopathic Physician*. "Without . . . viable osteopathic hospitals, the profession will become less important, as will [the] AOA, and jeopardize its future."[13] Bernhardt A. Zeiher, longtime CEO of Parkview Hospital in Toledo, echoed those sentiments when he was honored with the OOA Meritorious Service Award in 1992. In accepting the award, he told colleagues:

> The osteopathic profession has much for which to be thankful. The heritage to the present D.O. generation does not speak to the "blood, sweat and tears" that went into today's professional status, recognized educational programs and modern health care institutions. . . . Today's osteopathic profession must not fumble this position, must continue to pull itself up by its bootstraps, or [else] all of the past dedicated service is lost and the heritage will end. The osteopathic presence in northwest Ohio will only continue—and grow—with a strong osteopathic hospital, supported by loyal, dedicated osteopathic physicians providing quality medical care and education.[14]

The osteopathic hospital crisis was not confined to patient care. "If the hospitals are in trouble," OOA Executive Director Jon Wills wrote in his

At a meeting of the Ohio Osteopathic Hospital Association in Cleveland in 1968 are (*left to right*) Bernhardt A. Zeiher, OOHA president (representing Parkview Hospital), Harold A. Davis (Orrville Community Osteopathic Hospital), Florence Marstaller (Warren General Hospital), Ralph Keilch (Green Cross General Hospital), and John A. Rowland (Bay View Hospital). Zeiher, CEO of Parkview, frequently argued for the preservation of osteopathic hospitals to keep the profession strong. *Photo by Rebman Photo Service, Inc., courtesy of Ohio Osteopathic Association.*

1989 annual report, "how will we maintain quality osteopathic internships and residencies?"[15] Indeed, the issue of postdoctoral training presented a dilemma for the entire profession. Graduates of osteopathic colleges were increasing in number at the same time that the number of osteopathic hospitals was shrinking. Ohio's osteopathic postdoctoral training programs increasingly found themselves in fierce competition with M.D. institutions, which aggressively marketed their residency programs to D.O.'s, luring them away with more generous stipends and a wider variety of in-house specialty rotations.

Alarmed by the growing number of unfilled AOA-approved training slots, in March 1990 the OOA, OOHA, and OU-COM convened an education summit to explore ways to reverse this ominous trend. A working committee was appointed to address the issues in greater detail and report to the OOA House of Delegates. Robert J. George, D.O., director of medical education for Cuyahoga Falls General Hospital, chaired the committee, whose members included the Ohio Directors of Medical Education (DMEs), OU-COM administrators, regional assistant deans, and representatives of the OOA and OOHA. Richard A. Langsdorf, D.O., of Doctors Hospital in Massillon, was assigned to develop a means of linking the DMEs with the regional deans. A subcommittee consisting of OU-COM Dean Frank Myers, Mary L. Theodoras, D.O., Amelia G. Tunanidas, D.O., and John P. Sevastos, D.O., was charged with pursuing a closer affiliation between hospitals and the college. A second statewide summit, held in October 1990 at Doctors Hospital in Columbus, convened all Ohio residency trainers and DMEs to discuss the formation of a federation, or network, of all osteopathic training programs in Ohio. Its purpose would be to strengthen the quality of osteopathic postdoctoral education programs, including osteopathic training hospitals, and identify needs and priorities for osteopathic postdoctoral education in Ohio.[16]

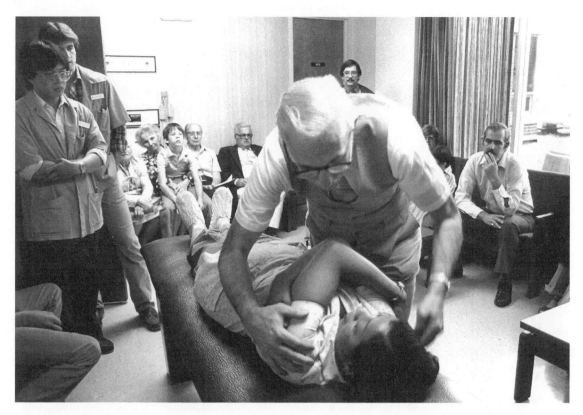

The education summits produced positive results. The Ohio DMEs formed their own organization and began meeting on a regular basis, with OOA Executive Director Jon Wills providing administrative assistance. They quickly adopted a position statement emphasizing teaching in internship programs; the traditional long hours devoted to patient care now would be balanced with a strong curriculum. They also determined to increase residency program stipends. The hospitals and OU-COM, meanwhile, agreed to give graduating Ohio residents a certificate bearing the college's name, and the OOA published an eighty-page reference book, *Osteopathic Postdoctoral Education in Ohio*, to better market Ohio's programs. Family practice residency program directors, meanwhile, formed an education consortium and began to network with other specialty programs, setting the stage for the more formalized statewide structure that would become the Centers for Osteopathic Research and Education (CORE).

In 1992, the OOA polled its members to determine what was on the minds of the state's D.O.'s. Reimbursement from their patients' insurance companies was the number one issue of concern, followed by the survival of osteopathic hospitals, osteopathic education, and government regulation.[17] In response to these findings, in 1993 the OOA Executive Committee voted to recommend restructuring the OOA Board and Executive

Willis W. Zimmerman, D.O., demonstrates osteopathic manipulative treatment (OMT) during an open house at Parkview Hospital in Toledo. Since the 1980s, OMT has enjoyed a resurgence of interest. *Courtesy of Ohio Osteopathic Association.*

Presenters at a forum titled "Women Physicians on Ethical Issues," held at Doctors Hospital (Columbus) in 1990, included many of Ohio's leading women doctors. *Left to right:* Elizabeth Connelly, D.O., chief of geriatric services at Akron General; Amelia Tunanidas, D.O., chief of staff at Youngstown Osteopathic Hospital; Jewell Malick, D.O., chair of the department of OB/GYN, Kirksville College; Marlene Wager, D.O., head of geriatric medicine, OU-COM; Mary L. Theodoras, D.O., regional assistant dean (Dayton), OU-COM; Anita Steinbergh, D.O., of Westerville; Suzanne Kimball, D.O., of Cleveland; Jean Drevenstedt, Ph.D., associate professor of psychology, Ohio University; and Ruth Purdy, D.O., staff internist, Doctors Hospital (Columbus). *Courtesy of Ohio Osteopathic Association.*

Committee to better represent the state's entire "osteopathic family." Representatives of OU-COM, the Ohio Osteopathic Hospital Association, and the Ohio Directors of Medical Education were added, reflecting the need for the profession and its institutions to work cooperatively to address the issues of foremost concern.

To address what might have been a devastating impact on postgraduate education nationwide, the AOA stepped into the breach by establishing, in 1995, a new method of evaluating and approving osteopathic postdoctoral training programs. That method, which required the accreditation of hospitals as Osteopathic Postdoctoral Training Institutions (OPTIs) in affiliation with an AOA-accredited college of osteopathic medicine, provided assurance that postgraduate programs, wherever they operated, would

Exhibitors at the 2000 OOA annual convention. From a small, almost negligible presence in the 1940s, exhibitors today fill an entire convention hall. *Courtesy of Ohio Osteopathic Association.*

meet or exceed established levels of quality. Collaborative efforts among OU-COM, CORE, OOA, and OOHA made Ohio the first state to meet the new standards, putting it—in the words of Fred Anthony, president of the Ohio Osteopathic Hospital Association—"a millenium ahead" of the rest of the nation in osteopathic post-doctoral education.[18]

In 1995, as the lack of managed-care contracts continued to erode the viability of freestanding osteopathic hospitals, the OOA called yet another summit, this time to explore ways of strengthening the profession's ability to compete. OOA, OU-COM, and OOHA leadership subsequently incorporated the Ohio Primary Health Care Alliance (OPHCA) to increase osteopathic bargaining power with managed-care organizations. The alliance was designed to "preserve and promote osteopathic education and its philosophy by providing high-quality and cost-effective health care in Ohio through a comprehensive and integrated health care delivery system."

OOA Legislative Director George Dunigan and Executive Director Jon F. Wills converse with Barbara Ross-Lee, D.O., dean of OU-COM, following the annual White Coat Ceremony, 1997. *Courtesy of Ohio University College of Osteopathic Medicine.*

OPHCA called for each osteopathic hospital and its medical staff to form its own physician hospital organization, or PHO. Each PHO would elect a physician and hospital representative to the board of OPHCA, which would seek managed-care contracts statewide.

Although some Ohio osteopathic hospitals did form local PHOs and join the osteopathic alliance, the focus for others shifted to self-preservation. Some hospitals felt compelled to put the OPHCA initiative on the back burner as they investigated merger or affiliation with larger hospital systems. At the same time, osteopathic medical staffs became confused. Should they buy a membership in OPHCA? Or should they join the larger, independent physician organizations (POs) then being formed to give individual doctors (D.O.'s and M.D.'s) greater leverage in dealing with hospitals and managed-care organizations alike? Conceived during the height of health care's evolution from fee-for-service to managed care, OPHCA, although visionary, was never fully implemented. "Our hospitals and physicians were unable to develop a common vision and respond quickly enough in light of the economic crisis that was developing at that time," OOA Executive Director Jon Wills would later recall.[19]

As the Ohio Osteopathic Association continued to work with the state's osteopathic hospitals and its member physicians to address the problems of managed care, mergers and acquisitions continued unabated. In June 1997, OOHA President Fred Anthony delivered a sobering report to the OOA House of Delegates. "Two of our osteopathic hospitals no longer exist: Warren General and Parkview (Toledo). The osteopathic physicians who practiced in those hospitals are still spinning and trying to find a comfort zone in which to practice—a place where they have some control of their future." In addition, Anthony reported, three other osteopathic hospitals had been merged into allopathic not-for-profit systems, and five others had been sold or were poised to be sold to allopathic for-profit systems.[20]

Indeed, the demise of Parkview Hospital in 1994 marked the beginning of the end of freestanding osteopathic hospitals in Ohio and contributed to growing feelings of uncertainty within the profession. Brentwood Hospital soon merged with the Meridia Health System, which itself was subsequently acquired by the Cleveland Clinic Health System. Columbia Healthcare Systems, a national for-profit corporation, acquired St. John West Shore Hospital; Quorum, Inc. acquired Doctors Hospital of Stark County; and MH Health Services, parent of St. Joseph Health Center, took over Warren General, reopening it as St. Joseph Health Center–Eastland. The "flagships," meanwhile—Doctors Hospital (Columbus) and Grandview

Hospital and Medical Center in Dayton—continued to look for potential business partners.

In 1995, Richmond Heights General Hospital was acquired by Primary Health Systems and became Mt. Sinai Medical Center–East. After courting a possible purchase by Columbia, Doctors Hospital sold its assets (North, West, and Nelsonville) to OhioHealth in 1998. And Grandview Hospital and Health Center, together with its smaller progeny, Southview Hospital and Family Health Center (opened as an outpatient facility in 1978), merged with Kettering Medical Center in 1999. Unable to recover from bankruptcy caused by declining admissions and revenue, Youngstown Osteopathic Hospital closed its doors on March 28, 2000. Finally, on October 1, 2001, Cuyahoga Falls General Hospital and Summa Health System signed an agreement to affiliate, leaving Selby General Hospital in Marietta as the only freestanding osteopathic hospital in Ohio.

In the midst of these losses, the profession was able to savor a notable victory. When Primary Health Systems filed for bankruptcy and closed Mt. Sinai–East (formerly Richmond Heights General Hospital) as part of a private deal negotiated with the Cleveland Clinic, the Cleveland Academy of Osteopathic Medicine rallied to keep the hospital open. In May 2000, U.S. Bankruptcy Judge Mary F. Walrath nullified the sale and ordered Mt. Sinai–East, together with Cleveland's St. Michael Hospital, to be sold by competitive bidding. Both hospitals were purchased by University Hospitals Health System of Cleveland, which agreed to operate them as full-service community hospitals. In 2003, University Hospitals Health System–Richmond Heights continued to operate as a teaching hospital affiliated with OU-COM.

"It has been a most difficult time for all of us," OOA President-elect Peter Bell, D.O., of Columbus, acknowledged in October 2001, as Doctors Hospital North was being converted to an outpatient and urgent-care facility. "Doctors Hospital North has been a stronghold of osteopathic practice, thought, politics, and training for over sixty years." Still, Bell stressed, "It is not, nor should it ever be, bricks and mortar that cement the bonds of our profession together. We are strong based on our conviction that osteopathic philosophy is the best way to practice medicine."[21]

There was a bright side to the shakeout. Many looked to hospital consolidation to improve health care by providing patients with the very best in technology. The nucleus of the osteopathic hospital system remained intact through CORE and continued affiliation with the Ohio Osteopathic Hospital Association. And didn't more mixed-staff hospitals, with a greater

The Osteopathic Difference
Donald C. Siehl, D.O.

As a physician, surgeon, teacher, author, and speaker, Donald C. Siehl, D.O. (1917–1994), of Dayton, was one of the osteopathic profession's most distinguished leaders. In 2002, OOA Executive Director Jon Wills recalled him as "a gentle giant, one of osteopathy's greatest defenders [who] couldn't talk about the profession without tears welling up in his eyes and the side of his mouth quivering from emotion."[1] A large-framed man, who sported dark-rimmed glasses and a crewcut, Don Siehl was born into an osteopathic family in Cincinnati. His father, Walter, was an osteopathic physician, as were an aunt, an uncle, and several cousins; four of his five brothers also followed him into the profession. "Each of us wanted to be our own boss and run our own business, so we all became doctors," he told the *Dayton Daily News* in 1978.

A 1939 graduate of Miami University in Oxford, Ohio, Don Siehl received his D.O. degree from Kirksville College of Osteopathic Medicine in 1943. He served his internship and orthopedic surgery residency at Doctors Hospital in Columbus. During World War II, he served as a second lieutenant in the U.S. Public Health Service. In 1951, Don Siehl established his practice in Dayton, where he was on the staff of Grandview Hospital. For thirty years, he served as director of the hospital's residency program in orthopedic surgery. At Grandview, he also served as chief of staff, chairman of the department of orthopedic surgery, and vice president of the board of trustees. Later in his career, Siehl became a part-time member of the clinical faculty of OU-COM. He was known as a demanding teacher and a fixture at commencement exercises, where he recited the osteopathic oath to graduating students.

Siehl's service to the profession was unstinting. He was president of both the OOA (1962–63) and the AOA (1978–79), president of the American College of Osteopathic Surgeons, a member of the American Osteopathic Board of Surgery for seventeen years, and executive director of the American Osteopathic Academy of Orthope-

dics. The recipient of many honors—including the OU-COM Phillips Medal of Public Service (1979), the AOA Distinguished Service Certificate (1984), and the OOA Distinguished Service Award (1990)—Siehl frequently urged his colleagues to use and promote the principles that made osteopathy a distinctive school of medicine, especially OMT. "The 'old timers' built the profession on this basis," he wrote in 1962. "We can keep it a strong profession chiefly on this basis."[2]

In 1978, at a meeting of the AOA House of Delegates in his hometown of Cincinnati, Siehl was installed as AOA president. For him, it was a sentimental occasion, as he revealed in his inaugural address. Siehl's words evidence his devotion to the osteopathic profession and the premium he placed on high-quality patient care.

> Forty years ago this week, in 1938, here in Cincinnati, I attended my first AOA convention as a pre-osteopathic student. My father, Dr. Walter H. Siehl, was local facilities chairman.

OOA President Donald J. Ulrich, D.O., passes the gavel to his successor, Donald C. Siehl, D.O., 1962. Siehl was among the most respected leaders of the modern profession. *Photo by Herb Topy Photo Service, courtesy of Ohio Osteopathic Association.*

Donald C. Siehl, D.O., received the Phillips Medal of Public Service in 1979 for osteopathic medical practice exemplifying the best tradition of family medicine. Named after Jody and J. Wallace Phillips, longtime benefactors of the university, the Phillips Medal is the highest honor bestowed by the Ohio University College of Osteopathic Medicine. *Left to right:* Frank W. Myers, D.O., then dean of OU-COM; U.S. Senator John Glenn, Jody Galbreath Phillips, Dr. Siehl, and State Representative Thomas Fries. Glenn and Fries were also honored with medals that year. *Courtesy of Ohio University College of Osteopathic Medicine.*

At that meeting Dr. Edward A. Ward of Michigan was AOA president. Dr. Arthur E. Allen of Minnesota was president-elect. Dr. Gertrud Helmecke [Reimer] of Cincinnati was first vice-president. Trustees included Dr. R. McFarlane Tilley, who also served as program chairman for the convention, and Dr. Paul T. Lloyd. New trustees were Dr. James O. Watson and Dr. John P. Wood.

Dr. Ira W. Drew was a member of the U.S. House of Representatives at the time and had just co-sponsored and helped pass a bill making D.O.'s eligible to take care of federal employees for [workers'] compensation injuries. The Advisory Board for Osteopathic Specialists was just being formed. The first breakfast meeting of the National Auxiliary was held, and the first scheduled meeting of the group that later became the American Academy of Osteopathy was held that week, under the aegis of Dr. Thomas L. Northup. As an interesting sidelight, a bugle call was used to summon delegates to the general session and section meetings.

It doesn't seem like forty years. I never dreamed then that I would be back in Cincinnati today, speaking to you as president of this great association.

My profession has been good to me. Our profession has been good to us. And we, individually and together, owe osteopathic medicine an obligation. That obligation isn't like a debt, something we can repay with money or by rendering a service. Rather, it is a responsibility, a life-long obligation, to *be* what we *are,* and to *do* what we are *supposed* to do.

What we are is osteopathic physicians and

surgeons. What we are supposed to do, in part, is take care of patients—not cases, not surgical procedures or diagnostic work-ups, but patients. And because we are *osteopathic* physicians, we are different, our approach *is* different, our aspirations are a bit higher.

We, as the osteopathic profession, are still trying to improve the practice of medicine, surgery, and obstetrics. Dr. Still started the first school of osteopathy for that purpose. Medical practice certainly is better today, but there still is room for improvement, there always will be.

We, as osteopathic physicians, have been taught that health care can be improved in very simple ways. We know that one secret of the care of the patient is caring for the patient. We know that showing patients that we care for them, as people, in *health* as well as in sickness, can be more powerful than any medication.

We know that no matter what the complaint, no matter how serious the disease or injury, patients have three primary needs—the need to be noticed, the need to be thought worthy, and the need to be needed, or loved. We can meet all those needs, fully, in the first moments, simply in the way in which we approach the patient. Our role, as Dr. Edward Trudeau has stated, is "to cure sometimes, to relieve often, and to comfort always."

Finally, we know that the character of the physician, the kind of person he is, is what patients notice first, react to most strongly, and remember longest. In fact, everything else being equal, it is the physician's character that most strongly determines his competence—his ability to do what he is supposed to do.

Let me remind you that the word "doctor" originally meant *teacher*. There are many similarities. The good doctor, like the good teacher, knows his subject, likes his subject, and likes and understands the people he deals with. Both the good doctor and the good teacher understand that much of what they do, and are supposed to do, is an art. This does not diminish the part of the doctor that is a scientist and a technician. Rather, it is through art that the science and technology of medicine become humanized, by the physician.

But doctors have teaching responsibilities beyond their patients. For us, particularly within the osteopathic profession, we have responsibilities to teach osteopathic students, interns, and residents. The clinical experience of a practicing D.O. simply cannot be duplicated, and that experience *must* be passed on, from generation to generation, in order to perpetuate the osteopathic difference. If we don't do it, nobody else will.

Finally, as practitioners of a learned profession, doctors have an obligation to teach the public generally. That responsibility too often is ignored in the daily rush of practice, but improving the public health must begin with improving the public's awareness.

So teaching, too, is what a doctor is supposed to do. But it doesn't end there. Physicians have other responsibilities—to their colleagues, to their hospitals, to their professional organizations, to their communities, and to their families. And let us not forget that the physician, ultimately, has a responsibility to himself—the obligation to fulfill his *own* needs, physical and emotional, so that these do not interfere with his primary obligation to his patients.

If we, each of us, do what we know we are supposed to do, most things will fall into place. That does not mean we will always have good results. It does not mean we will not make mistakes, have problems, avoid regrets.

What it does mean is a more balanced, less stressful, more productive and rewarding way of life. It means that we, in our most private moments, can know that we have done our very best—and no one can ask for more.

I, during the coming months, will be doing what I am supposed to do as your president—and I thank you, deeply and humbly, for this opportunity to serve the AOA, our great profession, and you.[3]

1. Jon F. Wills to the author, n.d. [2001].
2. "President's Message," *BOP,* July 1962, 3.
3. Inaugural Address, Cincinnati, Ohio, July 18, 1978, files of the OOA.

The Osteopathic Difference

John P. Sevastos, D.O.

"The D.O.'s are different," John P. Sevastos, D.O., says, "because of the way we train our students. The patient is not a 'disease entity,' he is a person." From his office in a small brick medical building near South Pointe Hospital in suburban Cleveland, Sevastos speaks of his profession with pride, not hesitating to draw a sharp distinction between allopathic and osteopathic medicine. M.D.'s, says Sevastos, represent "pure science"; D.O.'s "take a holistic approach."[1]

"When Mrs. Jones complains about a tightness in her chest, you listen attentively and you piece together your diagnosis, always remembering that Mrs. Jones has a family, she has responsibilities, [she has] stresses in life." If medical tests come back normal, he explains, "then you need to look at this patient as having some other underlying problem. You talk to her, you let her vent to you." That, says Sevastos, who is seventy-five, is the essence of holistic medicine and a defining feature of osteopathic medicine.

John Sevastos was born in 1926 in Woodville, Ohio, a small farm community located between Fremont and Toledo. Both of his parents emigrated from Greece. His mother was a homemaker; his father worked in an agricultural lime factory. Young John gravitated to medicine early on, when the only doctor in town—"an M.D. of tremendous stature and a fine gentleman"—took him under his wing. He tagged along on house calls, absorbing the lessons of his role model while the doctor, who had no sons, delighted in the boy's company.

In 1949, Sevastos was graduated from the University of Toledo College of Pharmacy, following which he worked briefly in the Cleveland Clinic pharmacy before taking a position as a medical service representative for Abbott Laboratories. Pharmacy, however, was only a means of financing his real dream: to become a doctor. Two experiences helped galvanize his decision to enter osteopathic medical school. One was an interview with the dean of Case Western Reserve University Medical School, who brusquely

John P. Sevastos, D.O., presents Joan McDonough, D.O., with the Northeast Ohio Regional Student Award, 1985. On the staff of Brentwood (now South Pointe) Hospital, Sevastos served as regional assistant dean of OU-COM and professor of family medicine and general practice from 1976 until 1994. *Photo by Denise M. Conrad, courtesy of Ohio University College of Osteopathic Medicine.*

discounted the value of Sevastos's pharmacy experience. The other was a brush with anti-osteopathic discrimination in his role as a pharmaceutical rep.

Sevastos at the time was ignorant of the distinction between D.O.'s and M.D.'s. One day the representative of another nationally known pharmaceutical company upbraided him for calling on a group of osteopathic physicians in the Osborn Medical Building in downtown Cleveland. "We're not allowed to call on those people," he was told. "They're quacks." To the contrary, Sevastos knew "those people" to be fine doctors and as knowledgeable about drugs as any M.D.'s he called on. One of them, James E. Coan, D.O., subsequently helped him enroll in the Chicago College of Osteopathy, from which he was graduated in 1956.

After interning at Forest Hill Hospital in Cleveland, Sevastos hung out his shingle in Warrensville Heights close by Brentwood Hospital, a new osteopathic facility built to serve the growing suburbs of southeastern Cuyahoga County. Several years later, he built a small brick medical building in the same neighborhood. At Brentwood (now South Pointe Hospital, part of the Cleveland Clinic Health System), Sevastos served as chairman of the department of general practice. From 1976 until 1994, he also served as a regional assistant dean of OU-COM and professor of family medicine and general practice.

Sevastos, an articulate man of regal bearing who favors opulent jewelry and large cufflinks, has been politically active in his profession. He has served as president of the American College of Osteopathic Family Physicians (1975–76), the Ohio Osteopathic Association (1981–82), and the American Osteopathic Association (1996–97). He is a charter member of the Ohio Osteopathic Political Action Committee, and in 1974 he was named National General Practitioner of the Year and Ohio General Practitioner of the Year. Avidly interested in osteopathic education, he has served on the board of the Pikeville (Kentucky) College of Osteopathic Medicine since its establishment in 1995.

In 2003, Sevastos was invited to deliver the A. T. Still Memorial Lecture at the annual meeting of the AOA House of Delegates—the highest honor that can be bestowed on an osteopathic physician. Speculating on how Andrew Taylor Still might react to the ten rules issued by the Institute of Medicine to guide patient-clinician relationships in the twenty-first century, he said, "As long as we as osteopathic physicians practice the type of medicine that was taught us in our osteopathic colleges, and we continue to treat our patients with compassion, understanding, and sincerity, we are in reality practicing twenty-first century medicine."[2]

In addition to its holistic approach, Sevastos believes that "the laying on of hands" is central to osteopathic medicine. He is a strong advocate of osteopathic manipulative treatment (OMT), calling it "an art . . . profound in its help in heal-ing." He recalls that during what he calls the "golden age" of osteopathic medicine—the 1950s and 1960s, when independent osteopathic hospitals flourished—"every post-op patient would receive OMT. Loosening up all of the joints, neck, spine . . . the circulation, the lymphatic draining. . . . You could just see the patients respond." But OMT takes time and effort, and its use declined with the rise of managed care; insurance companies would not reimburse doctors and hospitals for the extra care, a situation Sevastos calls "a tragedy." A decline in the number of OMT instructors and the increase in postgraduate training in allopathic hospitals also contributed to a decline in the practice of OMT. But Sevastos believes that OMT is enjoying a resurgence. "It's back now to where it was when I was a student"—thanks, he says, to prodding from "old weeds" such as himself, and to the American Osteopathic Association. In order to be accredited for osteopathic postgraduate training, the AOA requires that hospitals have an OMT department with a full-time OMT instructor.

Sevastos continues to preside over the same medical office he built in 1962. Today it is a family affair, staffed by Sevastos's son, Charles C. Sevastos, D.O. (also a Chicago graduate), and daughters Athena and Sherie, who manage the office. Asked if he is optimistic about the future of osteopathic medicine, Sevastos doesn't hesitate. "No question," he says decisively, at the same time allowing, "We are going to suffer some of the pain that all of medicine is suffering—and that is, there has been some decline in the applicant pool. Last year it was down across the board." That, he says, is because "the hardest thing in the world is to try to teach a young person holistic osteopathic medicine [when] you've got some insurance company with a clerk at the other end telling him what he can and can't do. This is creating some disillusion on the part of young people who might otherwise be interested in going into medicine. Managed care has hurt us a lot."

But Sevastos quickly banishes the gloomy thought. "Look at California [the 1961 merger of the state's D.O.'s with the California Medical

osteopathic presence, signal that D.O.'s had finally achieved equality with their M.D. counterparts? In hindsight, many viewed the avalanche of hospital mergers, acquisitions, and closures as testimony to the success and acceptance of the osteopathic profession itself. Richard A. Vincent, president and chief executive officer of the Osteopathic Heritage Foundations, would later say that osteopathic hospitals "were absolutely necessary at a point in the profession's history to provide osteopathic identity as well as facilities in which osteopathic physicians could practice. . . . However, as time passed and the profession gained recognition and [osteopathic] physicians were accepted into the mainstream, it was the osteopathic physicians [themselves] who demonstrated that these 'osteopathic' facilities were not as necessary as they once were."[22] Thanks to Ohio's osteopathic pioneers, doors that had once been closed were now open. In his 2000 inaugural address, OOA President Robert S. Juhasz, D.O., a member of the mixed-staff Cleveland Clinic, advised Ohio D.O.'s to "grow where you are planted"— be it at allopathic or mixed-staff health care institutions. "This has given us the opportunity and responsibility to share with our patients and colleagues the unique and distinctive nature of osteopathic medicine and [our] ability to offer treatments that others cannot," he said.[23]

Finally, propitiously, some hospital transactions negotiated with the profession's long-term viability in mind resulted in the creation of non-profit philanthropic foundations, funded by the sale of hospital assets, that promised to secure the future of osteopathic education, research, and patient care. The 1994 merger of Brentwood Hospital with Meridia Suburban Hospital, for example, resulted in the establishment of the Brentwood Foundation with an endowment of $14 million—money Brentwood had set aside in a "rainy day" fund. By June 2003, the Brentwood Foundation had

Foundations for the Future

Having weathered dizzying and often unsettling changes in American health care delivery, osteopathic medicine in Ohio is reinventing itself through an array of creative, patient-centered health care programs while playing a growing role in medical education and research. Leading the profession's transformation are the osteopathic foundations, which were endowed from the sales of osteopathic hospitals in Columbus, Cleveland, Dayton, and Massillon. Since 1999, the largest of these, the Columbus-based Osteopathic Heritage Foundations (OHF), has awarded more than $34 million for community health projects, medical education, and research.

One of the OHF's boldest research initiatives is under way in rural southeast Ohio. There, thanks to the $1.5 million endowment of the J. O. Watson, D.O. Research Chair, researchers and clinicians at the Ohio University College of Osteopathic Medicine (OU-COM) have established a diabetes treatment and research center as part of the college's Rural Health Institute. The center, under the direction of Leonard Kohn, M.D., formerly a director at the National Institutes of Health, builds on more than ten years of study on diabetes and the related issues of obesity and cardiovascular problems conducted by OU's Edison Biotechnology Institute. At the core of this effort are Goll-Ohio Eminent Scholar John Kopchick; Xiao Chen, associate professor of biomedical sciences; and Frank Schwartz, M.D., an endocrinologist who joined the faculty of OU-COM in 2003. "Because of our location and the expertise of our basic scientists and clinical faculty," says OU-COM Dean Jack Brose, D.O., "we have a chance to become a major center for diabetes research and treatment. Surveys have shown that Appalachian Ohio has one of the highest rates of diabetes in the nation, and nationally more money is spent each year on diabetes care than any other chronic disease."

Meanwhile, the Brentwood Foundation is leaving its mark on health care in northeast Ohio. In 2003, South Pointe Hospital in Warrensville Heights, a unit of the Cleveland Clinic Health System, unveiled a new $25 million west wing, including a twelve-suite surgery center, twenty-five-bed intensive-care unit, and outpatient testing center. Thanks to a $6 million grant from the Brentwood Foundation, the project also boasts a state-of-the-art medical education center, named after Brentwood Hospital founder Theodore F. Classen, D.O. It includes a medical library, auditorium, and classrooms; a skills lab allows doctors- and nurses-in-training to practice procedures before they see actual patients.

In southwest Ohio, Dayton's Grandview Medical Center, led by President Roy G. Chew, has won community praise by opening the Victor J. Cassano Health Center to serve the city's indigent population. Named for a local philanthropist and pizza entrepreneur, the $5.5 million health care and teaching facility, opened in 2003, combines the former family practice programs of the Hopeland Family Health Center and Corwin Nixon Community Health Center. Funded by grants from the federal government and supplemented by funds from the city of Dayton, the

Osteopathic residents observe a doctor-patient consultation at the Victor J. Cassano Health Center, opened in 2003 by Grandview Medical Center in Dayton. *Photo by Colin Gatland, courtesy of Grandview Medical Center.*

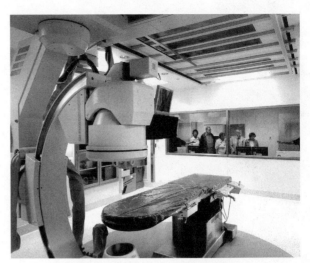

Local residents and elected officials tour South Pointe Hospital's new west wing during its dedication, September 21, 2003. The new high-tech facility, part of the former Brentwood Hospital, includes the Theodore F. Classen, D.O., Education Center, made possible by a $6 million grant from the Brentwood Foundation. *Courtesy of Ohio Osteopathic Association.*

Mathile Family Community Fund, and the Grandview Foundation, the Cassano Center offers comprehensive health care management and social services counseling in a state-of-the-art teaching facility.

Still other projects reflect a vital profession no longer inhibited by the bonds of prejudice but pioneering new paths. From a partnership with the Ohio Pharmacists Association and the Ohio Northern University Raabe School of Pharmacy to explore ways of reducing medication errors through electronic prescription, to OU-COM participation in a consortium for the study of traditional Chinese medicine, the osteopathic profession is exploring exciting new paths in its quest to improve the health, and health care, of Ohioans.

disbursed grants totaling more than $12.1 million for osteopathic education and research.[24] Similar foundations were created as a result of the merger of Grandview Hospital and Kettering Medical Center, the sale of Doctors Hospital of Massillon to Quorum Health Services, and the acquisition of Richmond Heights General Hospital by PHS, Inc. The acquisition of Doctors Hospital (Columbus) by OhioHealth added to the asset base of the existing Doctors Hospital Development Foundation, and the reorganized Osteopathic Heritage Foundations (OHF), with combined assets of $225 million, today constitutes the largest osteopathic health foundation in the country. Through August 2003, the OHF has awarded more than $34 million, with one-half that amount earmarked for the support of osteopathic medical education and research.[25]

In 1998–99, the Ohio Osteopathic Association marked its one-hundredth anniversary with a yearlong calendar of activities and special projects. Selecting the theme "A Distinguished Past—A Dynamic Future," a centennial celebration committee chaired by Richard Sims commissioned a video history and a poster. The OOA Board of Trustees invited its living past presidents to help develop new vision and mission statements to lead

Three of the six osteopathic physicians who have served on the State Medical Board of Ohio gather for a photograph with medical board Executive Director Thomas Dilling, October 2002. *Left to right (with their dates of service):* John E. Rauch, D.O. (1983–90), Theresa M. Hom, D.O. (1990–93), Mr. Dilling, and Anita M. Steinbergh, D.O. (1993–present). William J. Timmins Jr., D.O., who served on the medical board from 1972 until 1975, was absent due to illness. *Photo by Bill Pratt, courtesy of Ohio Osteopathic Association.*

Fifteen past presidents of OOA, whose tenures span nearly four decades, gathered for the 2003 annual convention. With them were longtime OOA Executive Director Jon F. Wills and AOA President Anthony A. Minnisale, D.O. *Front row, left to right:* Allen S. Birrer, D.O., Amelia G. Tunanidas, D.O., Peter A. Bell, D.O., Dr. Minnisale, Robert B. Black, D.O., and Gilbert S. Bucholz, D.O. *Back row, left to right:* Mr. Wills, John F. Uslick, D.O., M. Terrance Simon, D.O., E. Lee Foster, D.O., Robert S. Juhasz, D.O., Charles G. VonderEmbse, D.O., George Thomas, D.O., Robert J. Kromer, D.O., J. Richard Costin, D.O., David D. Goldberg, D.O., and John P. Sevastos, D.O. *Photo by Bill Pratt, courtesy of Ohio Osteopathic Association.*

OOA President E. Lee Foster, D.O., delivers the oath of office to incoming president Alison A. Clarey, D.O., of Dayton, at the 1999 annual convention. Closing the OOA's yearlong centennial celebration, Clarey urged the profession to remember that "we are the heirs of that first meeting [of osteopathic physicians] one hundred years ago." *Photo by Bill Pratt, courtesy of Ohio Osteopathic Association.*

the profession into the twenty-first century. At the close of the centennial year, OOA President Alison A. Clarey, D.O., of Dayton, took stock. "At the end of the 1990s," she observed in her inaugural address, "we are witnessing the merging and closing of our hospitals across the state, and a threat to our very existence. The threat is not just external, but also internal as we see a decline in the percentage of physicians joining local and state organizations." The disappearance of independent osteopathic hospitals, Clarey said, meant that the district academies "must play an increasingly important role as the focus of our interests." The academies—which had played so central a role in uniting Ohio D.O.'s at a time when they were denied access to hospitals and faced discrimination at every turn—now might once again serve to solidify professional bonds. "We dare not forget," she concluded, "that we are the heirs of that first meeting one hundred years ago. . . . The brick and mortar of the hospitals may be disappearing, but we are stronger in numbers than we have ever been. Our school is flourishing, and our way of practice is a standard for medical care."[26]

Osteopathic medicine in Ohio had made vast progress since thirteen pioneer physicians met in Mac and Adelaide Hulett's Columbus office that New Year's Eve day in 1898. In the space of one hundred years, the profession had achieved equal practice rights and representation on the state medical board. It had established excellent hospitals and a fine medical college. The concept of osteopathic medicine had expanded far beyond manipulation and surgery to embrace the use of pharmaceuticals and the diagnostic tools and therapeutic modalities of modern medicine. Osteopathic doctors now worked side-by-side with medical doctors in virtually every one of the state's hospitals, and they served on public health boards and commissions at every level of government. No longer a radical alternative, osteopathic medicine had moved ever closer to allopathic medicine to become a complete school of medicine—with something more to offer.

So what, one hundred years later, was the difference? At the heart of osteopathic medicine remained the core belief posited by Andrew Taylor Still in 1874: that osteopathic manipulation can restore normal musculoskeletal functions and thereby eliminate disease. Although the use of osteopathic manipulative treatment (OMT) had decreased for a time—one survey, in 1974, estimated that, of 53.5 million patient visits to office-based D.O.'s, fewer than 9.1 million (or 17 percent) included OMT[27]—by the mid-1980s there was a resurgence of interest. There was renewed emphasis on OMT in the colleges, and many longtime practitioners not only staunchly defended its efficacy but also preached its gospel to new generations of doctors. Even the medical establishment had begun to take note. Writing in the *New England Journal of Medicine* in 1999, researchers reported the results of a study comparing osteopathic spinal manipulation with standard care for lower back pain, concluding that while patients in both groups improved, the osteopathic-treatment group required significantly less medication.[28]

Osteopathic medicine could also claim a holistic, patient-centered approach focused on preventive care. And, if osteopathic physicians remained a distinct minority—in 2003, there were 3,090 licensed D.O.'s practicing in Ohio, compared with 24,535 licensed M.D.'s—they occupied a critical niche in the delivery of health care. For although D.O.'s represented just 12 percent of the total number of physicians in Ohio, they represented 26 percent of Ohio's family and general practice physicians.[29]

On its one-hundredth birthday, osteopathic medicine in Ohio continued to offer an important "second voice." In June 1999, as their predecessors had done a century earlier, Ohio's osteopathic physicians gathered in Columbus. At a centennial celebration gala at the Hyatt Regency Hotel, three hundred physicians devoted to the tenets of Andrew Taylor Still raised their Champagne glasses as E. Lee Foster, D.O., outgoing OOA president and master of ceremonies, offered a toast:

> To the distinctive and unique philosophy and practice of osteopathic medicine.
>
> To our hospitals, that gave us a place to practice and solidified our identity.

To our college, that added to our recognition and has become a beacon of educational excellence.

To the Ohio Osteopathic Association, the collective voice of Ohio D.O.'s who fought for our survival over the past century.

To the challenges and opportunities that will present themselves to us as we take our profession into the next century.

May those who follow us be as proud of our accomplishments as we are of those who came before us.

To our osteopathic family.

To our OOA.

To our dynamic future.[30]

Timeline of Osteopathic Medicine in Ohio

◌ = OHIO MILESTONE

1874   Frontier physician Andrew Taylor Still (1828–1917) first
        articulates basic osteopathic principles.

1892   First college of osteopathic medicine, the American
        School of Osteopathy (ASO), opens in Kirksville,
        Missouri.

1896   Vermont becomes the first state to license D.O.'s.

   ◌   Eugene Eastman, D.O., opens an office in Akron, becom-
        ing Ohio's first osteopathic physician.

1897   American Association for the Advancement of Osteopa-
        thy (renamed the American Osteopathic Association
        in 1901) founded in Kirksville, Missouri.

   ◌   Eugene Eastman, D.O., arrested in Akron for practicing
        medicine without a license.

   ◌   William J. Liffring, D.O., arrested in Toledo for practic-
        ing medicine without a license.

1898 ◌ Ohio Society for the Advancement of Osteopathy organ-
        ized on December 31 in Columbus.

        Associated Colleges of Osteopathy organized by repre-
        sentatives of legitimate schools.

1900 ◌ Ohio's Medical Practice Act amended, imposing impossi-
        ble educational requirements. The state's osteopathic
        physicians vow to defy the law.

   ◌   Hugh H. Gravett, D.O., arrested in Piqua.

1901   *Journal of the American Osteopathic Association* begins
        publication.

   ◌   Ohio Society for the Advancement of Osteopathy
        changes its name to the Ohio Osteopathic Society.

1902 ◌ Ohio Medical Practice Act amended to regulate osteopa-
        thy, creating a three-member osteopathic examining
        committee.

		First examination given for D.O. licensure in Ohio.
		AOA adopts standards for approval of osteopathic colleges. New AOA constitution provides that "graduates of those schools that are recognized by the association and no others shall be eligible for membership in this association."
1903		Emmons R. Booth, D.O., of Cincinnati, conducts first on-site inspection of osteopathic colleges.
		AOA seventh annual convention meets in Cleveland.
1904		AOA adopts code of ethics.
1905		Three-year course required for AOA-approved osteopathic colleges.
		Emmons R. Booth, D.O., of Cincinnati, publishes *History of Osteopathy, and Twentieth-Century Medical Practice.*
1906		AOA tenth annual convention meets in Put-in-Bay.
1910		Abraham Flexner, in report to Carnegie Foundation, exposes deficiencies in medical education and recommends that all but thirty-one medical schools be closed. "The country needs fewer and better doctors," he writes, "and the way to get them better is to produce fewer." He damns the eight osteopathic colleges, too, citing low entrance standards, poor basic science laboratories, insufficient clinical facilities, and inadequate teaching corps.
1911		Ohio Osteopathic Society publishes directory listing 102 members.
1912		First AOA headquarters office established in Orange, N.J.
1914		First osteopathic hospital in Ohio, Delaware Springs Sanitarium, opens.
1915		Four-year course required for AOA-approved osteopathic colleges.
1916		Ohio Osteopathic Society establishes legislative committee.
		Bulletin of the Ohio Osteopathic Society, edited by Frank A. Dilatush, D.O., begins publication.
1917		Ohio osteopathic physicians win right to use antiseptics and anesthetics and to perform minor and orthopedic surgery.

AOA sixteenth annual convention meets in Columbus.

1919   Major surgery added to subjects in which Ohio D.O.'s are to be examined by state medical board. In Section 1288 of General Code, word "Osteopath" is changed to "Osteopathic Physician."

1922   AOA headquarters office moves to Chicago.

1923   Ohio Osteopathic Society changes its name to Ohio Society of Osteopathic Physicians and Surgeons. There are six district organizations: Akron, Cincinnati, Cleveland, Columbus, Dayton, and Toledo.

The *Buckeye Osteopath* begins publication.

1926   Dayton Osteopathic Hospital opens.

1927   American Osteopathic Foundation is founded for support of research, publications, education, and other philanthropic activities.

Marietta Osteopathic Clinic opens.

1928   AOA establishes Distinguished Service Certificate awards.

1931   AOA establishes student loan fund and osteopathic Christmas seal program to raise money for student loans (later, student loans and research).

1934   Hinde Memorial Hospital opens in Sandusky.

1935   Gertrud Helmecke Reimer, D.O., of Cincinnati, becomes first woman president of Ohio Society of Osteopathic Physicians and Surgeons.

Cleveland Osteopathic Hospital opens.

1936   AOA conducts first inspection of institutions offering osteopathic internships.

The *Buckeye Osteopath*, renamed the *Buckeye Osteopathic Physician*, begins publication in November.

1938   Ohio Women's Osteopathic Association (later renamed the Auxiliary to the Ohio Osteopathic Association) organized.

1939   One year of preprofessional education required by AOA-accredited colleges.

In December, Ohio Society of Osteopathic Physicians and Surgeons hires William S. Konold as first executive secretary.

Ohio Osteopathic Hospital Association founded.

1940   Two years of preprofessional education required by AOA-accredited colleges.

Ohio Society of Osteopathic Physicians and Surgeons changes its name to Ohio Osteopathic Association of Physicians and Surgeons.

Doctors Hospital opens in Columbus.

Hayes-Mayberry Osteopathic Hospital opens in East Liverpool.

1943   Ohio D.O.'s win unlimited licensure and equal practice rights with M.D.'s.

J. O. Watson, D.O., of Columbus, appointed first osteopathic representative on state medical board.

AOA launches the Osteopathic Progress Fund. By close of first campaign, in 1944, raises nearly $1 million from practicing D.O.'s for support of osteopathic colleges.

Green Cross General Hospital opens in Akron.

1946   Andrew Taylor Still Memorial Lecture established (first lecture given in 1947).

Parkview Hospital opens in Toledo.

Mahoning Valley Green Cross Hospital opens in Warren.

Just over five hundred D.O.'s practicing in Ohio.

1947   AOA conducts first inspection of institutions offering osteopathic residencies.

Grandview Hospital opens in Dayton.

1948   New AOA headquarters building opens at 212 East Ohio Street, Chicago.

Bay View Hospital opens in Cleveland suburb of Bay Village.

1950   Forest Hill Hospital opens in Cleveland.

1953   Youngstown Osteopathic Hospital opens.

Sandusky Memorial Hospital opens.

Orrville Community Osteopathic Hospital opens.

1954   New Green Cross General Hospital opens in Akron suburb of Cuyahoga Falls (renamed Cuyahoga Falls General Hospital in 1978).

1956   President Dwight Eisenhower signs Public Law 763, permitting D.O.'s to be commissioned in medical corps of armed services.

OOA occupies new headquarters building at 53 West Third Avenue, Columbus.

1957	⬡	Brentwood Hospital opens in Cleveland suburb of Warrensville Heights.
1958		Three years of preprofessional education required by AOA-accredited colleges.
	⬡	Warren General Hospital opens.
1959	⬡	Northeastern Ohio General Hospital opens in North Madison.
1960	⬡	Doctors West opens in Columbus suburb of Lincoln Village.
	⬡	Otto C. Epp Memorial Hospital opens in Cincinnati.
1961	⬡	Richmond Heights General Hospital opens.
		California Osteopathic Association votes to merge with California Medical Society.
1962		California voters approve proposition removing power of Board of Osteopathic Examiners to license new D.O.'s. Some twenty-five hundred California D.O.'s receive M.D. certificates. College of Osteopathic Physicians and Surgeons, Los Angeles, is converted to allopathic medical college (University of California–Irvine).
1963		U.S. Civil Service Commission announces that, for its purposes, M.D. and D.O. degrees would henceforth be considered equivalent.
	⬡	New East Liverpool Osteopathic Hospital opens.
	⬡	Doctors Hospital of Stark County opens in Massillon.
1965	⬡	New Selby General Hospital opens in Marietta.
1966		Secretary of Defense Robert S. McNamara orders all armed services to accept qualified D.O.'s as military physicians and surgeons.
		AOA accepted as accrediting agency for determining osteopathic hospitals' eligibility for participation in Medicare program.
1967		First D.O.'s drafted as medical officers in armed forces.
1968		AMA urges county and state medical societies to open their memberships to D.O.'s.
	⬡	William S. Konold resigns as OOA executive director. Richard L. Sims named to succeed him.
	⬡	OOA-commissioned study by Ohio State University researchers finds public is largely uninformed about D.O.'s.

		Ohio Osteopathic Medical Assistants Association founded.
1970		AOA gives approval for D.O.'s to enter allopathic residencies.
		Michigan State University College of Osteopathic Medicine, the profession's first university-affiliated osteopathic medical college, opens.
1972		Commemorative postage stamp honors seventy-fifth anniversary of AOA.
		OOA House of Delegates passes resolution urging study of feasibility of creating an osteopathic medical school in Ohio.
1973		Mississippi, last of "limited practice" states, grants full practice rights to licensed D.O.'s.
1974		Centennial of osteopathy observed.
		OOA hires Christian H. Kindsvatter as executive director.
		OOA launches college campaign.
		California State Supreme Court rules that licensing of D.O.'s in state must be resumed.
1975		Gov. James A. Rhodes signs bill creating Ohio University College of Osteopathic Medicine (OU-COM) in Athens.
		Evelyn Cover, D.O., of Columbus, is appointed to state medical board.
1976		In September, OU-COM opens with first class of twenty-four students.
1977		OOA names Jon F. Wills, director of public relations, as executive director.
		AOA approves postgraduate education for joint-staff hospitals willing to apply for AOA accreditation.
1979		More than one thousand new D.O.'s graduate.
		Just over fourteen hundred D.O.'s are practicing in Ohio.
1980		OU-COM graduates first class.
		Family Health, syndicated radio program produced at OU-COM, debuts.
1981		Bay View Hospital closes after contracting with St. John Hospital to build new joint-staff St. John West Shore Hospital.
1982		More than twenty thousand D.O.'s are practicing nationwide.

		OU-COM's entering class of one hundred represents full enrollment.
1984		OOA launches multimedia campaign in selected markets to boost public awareness of D.O.'s.
1987		AOA purchases and occupies new headquarters building at 142 East Ontario Street, Chicago.
1988		Northeastern Ohio General Hospital in Madison closes.
1991		OU-COM alumni comprise 10 percent of practicing D.O.'s in Ohio.
1992		Osteopathic education sets new record with just over two thousand freshmen enrolled in nation's fifteen AOA-accredited colleges.
1994		Parkview Hospital in Toledo closes.
		Brentwood Hospital in Warrensville Heights merges with Meridia Health System.
1995		AOA establishes new method for accrediting hospitals as Osteopathic Postdoctoral Training Institutions (OPTIs).
		Ohio becomes first state to meet new OPTI standards.
1997		AOA celebrates its centennial.
1998		American Medical Student Association Foundation ranks OU-COM first in nation for producing family doctors.
		Doctors Hospital (Columbus) acquired by OhioHealth.
1998		OOA celebrates its centennial.
2000		Youngstown Osteopathic Hospital closes.
2003		Cuyahoga Falls General Hospital acquired by Summa Health System.
		Through 2003, Brentwood Foundation disburses grants of more than $12 million; Columbus-based Osteopathic Heritage Foundations, more than $34 million.
		Just over three thousand D.O.'s are practicing in Ohio.

APPENDIX 2

Past Presidents of the Ohio Osteopathic Association

1898–1899	George W. Sommer, D.O.
1899–1900	Hugh H. Gravett, D.O.
1900–1901	Hugh H. Gravett, D.O.
1901–1902	Clarence V. Kerr, D.O.
1902–1903	DeWitt C. Westfall, D.O.
1903–1904	John F. Bumpus, D.O.
1904–1905	Oliver G. Stout, D.O.
1905–1906	Edward W. Sackett, D.O.
1906–1907	Hiram E. Worstell, D.O.
1907–1908	Mac F. Hulett, D.O.
1908–1909	Charles L. Marsteller, D.O.
1909–1910	Albert W. Cloud, D.O.
1910–1911	Emmons R. Booth, D.O.*
1911–1912	Louis C. Sorenson, D.O.
1912–1913	Allen Z. Prescott, D.O.
1913–1914	Roy W. Sanborn, D.O.
1914–1915	Roy W. Sanborn, D.O.
1915–1916	William A. Gravett, D.O.*
1916–1917	Charles A. Ross, D.O.
1917–1918	Earl H. Cosner, D.O.
1918–1919	Earl H. Cosner, D.O.
1919–1920	Percy E. Roscoe, D.O.
1920–1921	Eugene E. Ruby, D.O.
1921–1922	Reginald H. Singleton, D.O.
1922–1923	Reginald H. Singleton, D.O.
1923–1924	Richard A. Sheppard, D.O.
1924–1925	Bert C. Maxwell, D.O.
1925–1926	Nicholas A. Ulrich, D.O.
1926–1927	Ralph P. Baker, D.O.
1927–1928	H. Lynn Knapp, D.O.
1928–1929	H. Lynn Knapp, D.O.

1929–1930	Herman L. Samblanet, D.O.
1930–1931	Herman L. Samblanet, D.O.
1931–1932	Meryl A. Prudden, D.O.
1932–1933	Arthur E. Best, D.O.
1933–1934	Eugene C. Waters, D.O.
1934–1935	Harold J. Long, D.O.
1935–1936	Gertrud Helmecke Reimer, D.O.
1936–1937	Richard A. Sheppard, D.O.
1937–1938	Edgar H. Westfall, D.O.
1938–1939	Hubert L. Benedict, D.O.
1939–1940	Ralph S. Licklider, D.O.
1940–1941	Ralph S. Licklider, D.O.
1941–1942	Donald V. Hampton, D.O.*
1942–1943	Donald V. Hampton, D.O.
1943–1944	John W. Mulford, D.O.*
1944–1945	Homer R. Sprague, D.O.
1945–1946	Warren G. Bradford, D.O.
1946–1947	Charles F. Rauch, D.O.
1947–1948	Charles L. Ballinger, D.O.
1948–1949	Walter H. Siehl, D.O.
1949–1950	Robert F. Haas, D.O.
1950–1951	Theodore C. Hobbs, D.O.
1951–1952	Domenic J. Aveni, D.O.
1952–1953	Roger E. Bennett, D.O.
1953–1954	Charles L. Naylor, D.O.*
1954–1955	W. Dayton Henceroth, D.O.
1955–1956	William B. Carnegie, D.O.
1956–1957	John W. Hayes, D.O.*
1957–1958	Jack M. Wright, D.O.
1958–1959	Robert L. Thomas, D.O.
1959–1960	A. R. Fuller, D.O.
1960–1961	Leonard D. Sells, D.O.
1961–1962	Donald J. Ulrich, D.O.
1962–1963	Donald Siehl, D.O.*
1963–1964	Robert H. Sheldon, D.O.
1964–1965	James E. Walker, D.O.
1965–1966	Jack D. Hutchison, D.O.
1966–1967	Robert J. Kromer, D.O.
1967–1968	George S. Cozma, D.O.

1968–1969	Robert G. Neth, D.O.
1969–1970	Eugene R. DeLucia, D.O.
1970–1971	Allyn W. Conway, D.O.
1971–1972	Layton S. Shaffer, D.O.
1972–1973	Samuel W. Howe, D.O.
1973–1974	J. Richard Costin, D.O.
1974–1975	J. Arnold Finer, D.O.
1975–1976	Martin E. Levitt, D.O.
1976–1977	James C. Ward, D.O.
1977–1978	Philip Golding, D.O.
1978–1979	Gilbert S. Bucholz, D.O.*
1979–1980	Donald L. Turner, D.O.
1980–1981	Allen S. Birrer, D.O.
1981–1982	John P. Sevastos, D.O.*
1982–1983	Duane J. Kerscher, D.O.
1983–1984	Carmin S. Maietta, D.O.
1984–1985	Mary L. Theodoras, D.O.
1985–1986	Andre V. Gibaldi, D.O.
1986–1987	William J. Stefanich, D.O.
1987–1988	E. Thomas Harnish, D.O.
1988–1989	Robert B. Black, D.O.
1989–1990	George Thomas, D.O.*
1990–1991	John F. Uslick, D.O.
1991–1992	Charles G. VonderEmbse, D.O.
1992–1993	Rolan J. Bingham, D.O.
1993–1994	David D. Goldberg, D.O.
1994–1995	Emil E. Pogorelec, D.O.
1995–1996	Eugene D. Pogorelec, D.O.
1996–1997	Carl R. Backes, D.O.
1997–1998	Amelia G. Tunanidas, D.O.
1998–1999	E. Lee Foster, D.O.
1999–2000	Alison A. Clarey, D.O.
2000–2001	Robert S. Juhasz, D.O.
2001–2002	M. Terrance Simon, D.O.
2002–2003	Peter A. Bell, D.O.
2003–2004	Paul A. Martin, D.O.

*Also served as president of the American Osteopathic Association.

Ohio's Osteopathic Hospitals

One measure of the Ohio profession's progress, especially after World War II, was the establishment of osteopathic hospitals. From six in 1940, the number grew to seventeen by 1965, when Selby General Hospital in Marietta was dedicated. Following this heady growth, many hospitals closed or merged with larger allopathic hospital systems due to rising costs, a more competitive health care environment, and changes in Medicare reimbursement. Irrespective of size, these institutions played a notable role in the delivery of health care in their communities and established Ohio as a national leader in the training of osteopathic physicians.

BAY VIEW HOSPITAL, BAY VILLAGE

Bay View Hospital traces its origin to the Cleveland Osteopathic Hospital, established in 1935 by a group of doctors led by Richard A. Sheppard, D.O., then head of surgery at the downtown Cleveland Osteopathic Clinic. In 1947, the Cleveland Osteopathic Hospital acquired a twenty-four-room mansion on Lake Erie in the West Side Cleveland suburb of Bay Village. Built by Cleveland industrialist Washington H. Lawrence in 1898, the three-story mansion with its large rooms and fireproof construction was readily adapted to hospital use. Ohio Gov. Thomas J. Herbert was among the dignitaries at the opening of the new sixty-bed Bay View Hospital on Sunday, October 3, 1948. (The Cleveland Osteopathic Hospital was continued for a time as a clinic.) In 1953, Bay View added a new wing, with Hill-Burton funds contributing one-third of the $385,000 cost. The addition increased capacity to eighty-three beds and twenty bassinets, and added twenty-four-hour emergency service. A decade later, two floors, each housing nursing units of thirty beds, were added atop the 1953 wing.

In 1970, Bay View's application for further expansion was turned down by the Metropolitan Health Planning Corporation, which ordered Bay View to close and combine its services with one or more West Side hospitals. In 1978, Bay View Hospital and St. John Hospital, under the sponsorship of the Sisters of Charity of St. Augustine, formed a partnership to build a new joint-staff facility. The two-hundred-bed, acute-care St. John West Shore Hospital opened its doors in 1981. It was the first hospital in the United States to merge osteopathic and allopathic institutions under one roof, with co-chairs of specialty departments and equal representation on the executive committee. Jack J. Brill, D.O., who had practiced family medicine at Bay View for almost thirty years, served as vice president of the medical staff and co-chairman of the family practice department. As the hospital's first director of medical education, he fought to gain approval for an osteopathic family practice internship. The hospital accepted its first six osteopathic students in July 1999, two months before his death.

BRENTWOOD HOSPITAL, WARRENSVILLE HEIGHTS

Opened in January 1957 as a sixty-bed acute-care general hospital, Brentwood Hospital was the project of Theodore F. Classen, D.O., who became known as a "one-man army" in the battle to build a hospital to serve the fast-growing southeastern part of Cuyahoga County. Classen, a graduate

of the Kansas City College of Osteopathy and Surgery, formed a private corporation and began selling stock while he continued his surgical practice. With $350,000 in commitments—much of it from osteopathic patients of limited means—he obtained a $480,000 bank loan to complete the project. Classen served as chief of staff; Paul J. Stitzel was administrator; and Leonard C. Nagel, D.O., was chief of orthopedics prior to his untimely death, in 1963, at age fifty-nine.

A new $550,000 wing completed in 1963 increased bed capacity to ninety-three. Staff doctors and trustees gave more than half the total needed, and other gifts came from businesses and founda-

Breaking ground for Brentwood Hospital, 1956. Opened in 1957, Brentwood was the realization of a dream for Theodore F. Classen, D.O. (*left*), who was determined to bring osteopathic hospital service to the expanding suburbs of southeastern Cuyahoga County. *Courtesy of South Pointe Hospital.*

tions. Four years later, in October 1967, Ohio Senator Frank Lausche was the featured speaker when Brentwood dedicated a new $2.9 million addition, bringing capacity to 150 beds.

In 1979, the Brentwood Ambulatory Care Center was dedicated in Sagamore Hills to serve a fast-growing area of Summit County. "Everything Brentwood has is here also, except beds," Classen said of the $1.5 million emergency and outpatient facility. He added, "Our concern is to keep patients out of bed, because of the great cost involved." Ted Classen—born Ted Klasinski—never forgot his roots, and in 1984 Cleveland's Slavic Village neighborhood celebrated the opening of the Brentwood Family Health Center at 6511 Fleet Avenue.

Brentwood was the first osteopathic hospital in Ohio approved for the training of residents in family practice (1973) and the first to implement a paramedic-training program certified by the State of Ohio (1976). But the hospital's exclusion as a preferred provider by many health insurance plans and a declining patient census caused Brentwood to merge with the four-hospital Meridia Health System in 1994. Two years later, the Cleveland Clinic Health System acquired Meridia.

Osteopathic practice and medical education continue at the former Brentwood Hospital, now known as South Pointe Hospital. The Brentwood Foundation, created at the time of the 1994 merger, supports projects related to osteopathic health care and education, especially at South Pointe Hospital and in the local community. With assets (2002) of $20 million, the foundation's recent projects include a mobile wellness clinic serving Warrensville Heights public schools and the establishment of the Center of Excellence for Osteopathic Medical Education. The latter project includes construction of a $6 million medical education facility at South Pointe for the support of medical research and publication, faculty development, and physician and faculty recruitment. Terri Kovach has served as executive director of the Brentwood Foundation since 1996.

CUYAHOGA FALLS GENERAL HOSPITAL, CUYAHOGA FALLS

Cuyahoga Falls General Hospital traces its roots to 1943, when a small group of osteopathic physicians and surgeons in Akron purchased and renovated a building at 15 Broad Street that had formerly housed the East Akron Community Hospital. The thirty-five-bed, nonprofit Green Cross General Hospital opened its doors on September 20 that year, with officers Arthur L. Harbarger, D.O., president; E. L. Jay, D.O., and L. O. Wilkins, D.O.,

Administrator Ralph F. Keilch, Lloyd O. Wilkins, D.O., and Arthur L. Harbarger, D.O., review expansion plans for Green Cross General Hospital in Cuyahoga Falls, 1965. *Courtesy of Ohio Osteopathic Association.*

vice presidents; and H. S. Jeffers, D.O., secretary-treasurer. William S. Konold served as consultant-manager. A decade later, ground was broken for a new hospital at 1900 23rd Street in nearby Cuyahoga Falls. The new seventy-eight-bed Green Cross General Hospital, financed in part by $200,000 in Hill-Burton funds, was formally opened to the public on December 4, 1954. The hospital experienced steady growth, and by 1968—its twenty-fifth anniversary—Green Cross boasted accommodations for 275 patients and extensive surgery, radiology, laboratory, and medical education facilities. An ambulatory-care center was added in 1975. In 1978, the hospital changed its name to Cuyahoga Falls General Hospital to better identify with the area it served, and announced the opening of a family practice center. The full-service, primary-care hospital remained a freestanding osteopathic hospital until 2001, when it became part of the Summa Health System.

DOCTORS HOSPITAL, COLUMBUS

Doctors Hospital was founded in 1940 by a group of osteopathic physicians and surgeons headed by Ralph S. Licklider, D.O., James O. Watson, D.O., and Harold E. Clybourne, D.O. They purchased Columbus Radium Hospital, which occupied a large Second Empire–style mansion on the city's near North Side, and extensively remodeled it. The new hospital, with twenty-four beds, seven bassinets, and a staff of twenty-one doctors, opened on August 15, 1940. A new north wing completed in 1945 brought capacity to sixty-four beds, but within months demand outpaced supply. A new surgery suite was opened in 1949, a new south wing in 1950, and yet another wing in 1956. Doctors Hospital now boasted 225 beds, six operating rooms, and extensive auxiliary and support facilities, including its own quarters for nurses, interns, and residents.

In October 1963, with twenty-five hundred people looking on, Doctors Hospital dedicated a new 112-bed facility in the Columbus suburb of Lincoln Village. Management, purchasing, billing, laundry, and the more sophisticated diagnostic procedures were centralized at Doctors North, while microwave transmitters allowed educational programs to be shared. The

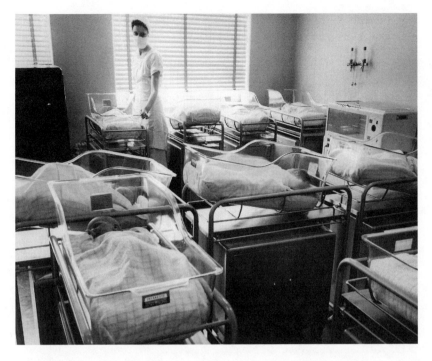

The baby boom was under way when this photograph was taken at Doctors Hospital in Columbus, about 1955. *Photo by Howell Associates, courtesy of Ohio Osteopathic Association.*

Buckeye hailed Doctors West, the nation's first "satellite" hospital, as "the most unique system of hospital care in the country today."

Doctors' growth continued with its 1980 acquisition of Mount St. Mary Hospital in Nelsonville, near the new Ohio University College of Osteopathic Medicine, which it renamed Doctors Hospital of Nelsonville. Successive expansion projects, meanwhile, brought bed capacity at Doctors North to 323. By 1990, with a staff of more than 450 physicians and 700 allied health care professionals, Doctors had emerged as one of the largest osteopathic teaching hospitals in the United States and a cornerstone of osteopathic medicine in Ohio.

In 1997, facing a highly competitive managed-care environment, the hospital began to consider possible acquisition partners. The following year, Doctors agreed to sell its assets to OhioHealth, the nonprofit parent of two other Columbus hospitals, Grant Medical Center and Riverside Methodist Hospital. The acquisition agreement provided for the continuation of osteopathic postgraduate training, continued partnership with CORE, equal recognition of osteopathic physicians within all OhioHealth facilities, and AOA accreditation. Proceeds from the sale further enhanced the asset base of the Osteopathic Heritage Foundations (OHF), with its mission "to improve the health and quality of life in the community through education, research and service consistent with our osteopathic heritage." Through 2002, OHF had directed over $24 million toward the endowment of osteopathic postgraduate education at Doctors OhioHealth, the support of a community health access project in underserved areas of central Ohio, an oral health initiative in southeast Ohio, and osteopathic research and education, including endowed chairs at OU-COM and the Texas College of Osteopathic Medicine. Richard A. Vincent was president and chief executive officer of OHF, whose combined assets exceeded $225 million, making it one of the largest health foundations in Ohio and the largest osteopathic health foundation in the United States.

DOCTORS HOSPITAL OF STARK COUNTY, MASSILLON

Doctors Hospital of Stark County was conceived and planned by a group of osteopathic physicians with large, well-established practices who, at the time, were denied staff privileges at local hospitals. They were Isadore Browarsky, Irving M. Gordon, E. William Kenneweg, Thomas E. Violand, and Joseph F. Rader, all D.O.'s of Massillon, and Herman L. Samblanet and G. E. Brooker, D.O.'s of Canton. Surgeon J. Arnold Finer, D.O., of Youngstown, who was seeking to relocate, joined the project after answering a journal ad placed by the group. The fifty-bed nonprofit community hospital, located in Perry Township midway between Canton and Massillon, was dedicated on December 15, 1963. Two years later, the one-story brick hospital was doubled in capacity, to one hundred beds.

Doctors Hospital grew rapidly, reflecting Stark County's population growth. In 1971, a $2.5 million expansion added eleven intensive-care beds, three operating rooms, a new clinical laboratory and central supply, expanded recovery rooms, and a relocated emergency area. In 1980, Doctors Hospital completed a $4.1 million expansion and renovation project, including a new two-level building to accommodate the medical records department, physical therapy, nursing education, and medical education library. In 1994, the hospital broke ground for a three-story, $10.5 million addition and renovation.

Today Doctors Hospital hosts forty-seven (intern and resident) postgraduate positions and is affiliated with nurse education programs at several Ohio colleges and universities. An unusual aspect of the hospital's history is its longstanding service to the Amish and Mennonite communities of Holmes, Tuscarawas, and Wayne Counties. Family members of patients may stay free of charge at a hospital-owned residence nearby, and Doctors provides special accommodations for horses and buggies.

In 1996, the hospital was acquired by a joint partnership formed by Quorum Health Group, Summa Health System, and the Cleveland Clinic

Groundbreaking ceremonies for Doctors Hospital of Stark County, May 6, 1963, with four of its founders wielding shovels. *Left to right:* Isadore Browarsky, D.O., Thomas E. Violand, D.O., Irving M. Gordon, D.O., and E. William Kenneweg, D.O. *Courtesy of Ohio Osteopathic Association.*

Foundation. The deal resulted in formation of the Austin-Bailey Health and Wellness Foundation, with a mission to support programs that promote the physical and mental well-being of the people of Holmes, Stark, Tuscarawas, and Wayne Counties. Assets in 2002 totaled just over $8 million. Don A. Sultzbach was executive director. In 2001, Quorum was purchased by Triad Hospitals, with which Doctors Hospital is now affiliated.

EAST LIVERPOOL OSTEOPATHIC

HOSPITAL, EAST LIVERPOOL

East Liverpool Osteopathic Hospital traces its roots to the Hayes-Mayberry Osteopathic Hospital, started in 1940 as a surgical facility for John W. Hayes, D.O., a general surgeon, and C. M. Mayberry, D.O., an eye-ear-nose-throat specialist. The two doctors began modestly, with three beds, shortly increased to six, on the second floor of an office building at 142 West Fifth Street. As osteopathic physicians from the surrounding area used the hospital, the limited space proved inadequate. At a reorganization meeting on January 31, 1946, in the Steubenville office of T. O. Rogers, D.O., new

Open house and dedication of East Liverpool Osteopathic Hospital, April 1963. *Courtesy of Ohio Osteopathic Association.*

officers and trustees were elected. The hospital purchased property at 203 and 205 West Fifth Street, East Liverpool. There, a large two-and-one-half-story Queen Anne–style house was converted to hospital use, bringing bed capacity to twenty-nine. In the late 1940s, the hospital changed its name to East Liverpool Osteopathic Hospital.

In 1960, the hospital launched a public drive for funds to build a new osteopathic hospital. Patients of the hospital's physicians, members of the hospital staff, and local industries contributed generously, with one-third of the amount for the new hospital coming from Hill-Burton funds. The new forty-bed hospital, which incorporated the existing house, was dedicated on Sunday, April 28, 1963, with the citizens of East Liverpool attending an open house and tour of the new facilities.

Following the retirement of the two doctors who had spearheaded its establishment and growth, the hospital was unable to attract additional osteopathic physicians. It was subsequently purchased by three allopathic physicians, who reopened the hospital in 1977 as Potter's Medical Center, a psychiatric and neurosurgical facility.

OTTO C. EPP MEMORIAL HOSPITAL, CINCINNATI

Construction of Cincinnati's Otto C. Epp Memorial Hospital began in December 1958, thanks to the generous gift of $500,000 from a local businessman. The forty-eight-bed facility opened in February 1960, filling a longstanding need for osteopathic hospital service in Cincinnati. William S. Konold equipped the hospital and served as administrator. At the dedication ceremonies, Clara Wernicke, D.O., a student of Andrew Taylor Still who had practiced in Cincinnati since 1901, received special honors.

In 1967, a $1 million addition ($370,000 of this amount in Hill-Burton funds) enlarged Epp Memorial to eighty-four beds. In 1985, as Epp Memorial marked its twenty-fifth anniversary, the hospital broke ground for a new $1 million emergency department. It was named after Stephen J. Thiel, D.O., a hospital founder who had served as medical director and chairman of the hospital board.

In 1988, Otto C. Epp Memorial Hospital was acquired by Midwest Health Systems, parent of Cincinnati's Jewish Hospital. The agreement called for Epp Memorial to retain its own medical staff and board, and to pursue accreditation by the Joint Commission on Accreditation of Healthcare Organizations (JCAHO) in return for Midwest's commitment to make extensive capital improvements. However, as Epp Memorial underwent major renovation and added new medical services, it quickly lost its osteopathic identity and was renamed the Jewish Hospital Kenwood. Following the closure of Jewish Hospital's Burnet Avenue facility in 1997, it became the main campus of the Jewish Hospital.

GRANDVIEW HOSPITAL, DAYTON

Drs. Heber M. Dill, Frank A. Dilatush, and William A. Gravett established the Dayton Osteopathic Hospital in 1925. One of Ohio's first osteopathic hospitals, the ten-bed facility occupied a former private residence at 325 West Second Street, near the downtown business district. Drs. Dill, Dilatush, and Richard F. Dobeleit, a radiologist and surgeon, reorganized the hospital in 1935, and in 1940 it was incorporated as a nonprofit institution.

In 1947, the hospital moved to a new building on Grand Avenue and was renamed Grandview Hospital. Grandview added an eighty-bed, $650,000 addition in 1952 and a $750,000 maternity wing in 1958. That year the hospital admitted 11,000 patients and counted 4,860 surgeries and 1,725 births. It also established a new division of nuclear medicine under the direction of J. Milton Zimmerman, D.O., touted as "the first major isotope laboratory in an osteopathic hospital, for both diagnosis and therapy." Expansion continued, with a sixty-four-bed addition in 1962—"The extra capacity will relieve the waiting problem," the *Buckeye* observed, "but it will not eliminate it"—and a $5.5 million expansion and remodeling program completed in 1970.

In 1978, Grandview opened a new $12.5 million Ambulatory Care Center (ACC) in Washington Township, near the Dayton Mall. With two acres under roof, the one-story building provided twenty-four-hour outpatient services. In 1982, Grandview broke ground for two ambitious expansion projects: a fifty-six-bed inpatient addition to the ACC; and the expansion and renovation of Grandview Hospital, including a new main entrance and admitting area, emergency room, surgery suites, outpatient clinic, and intensive and critical care areas. The ACC was renamed Southview Hospital and Family Health Center, and Grandview Hospital became Grandview Hospital and Medical Center. By 1990, the combined facilities, with 452 beds, offered comprehensive health care services, including acute medical care, emergency care, maternity care, mental health services, open-heart surgery, and magnetic resonance imaging. Together forming a regional teaching center for OU-COM, the two hospitals comprised the nation's third-largest osteopathic health system, with more than nineteen hundred employees and three

Grandview Hospital and Medical Center, Dayton, 1997. *Courtesy of Ohio Osteopathic Association.*

hundred staff physicians. In 1993, Huber Health Center joined the system as an urgent care facility for communities north of Dayton.

In 1999, in response to the challenges of a highly competitive managed-care environment, Grandview and Southview Hospitals merged with Kettering Medical Center, operated by Seventh-day Adventist Health Care Systems. The merger, hailed by an editorial in the *Dayton Daily News* (April 14, 1999) as "a progressive blend of allopathic and osteopathic medicine that ought to be a national trend," positioned Grandview to undertake new initiatives. When Dayton's Franciscan Medical Center abruptly closed its doors in 2000, Grandview absorbed its adult psychiatric program, kept its large health clinic open, and salvaged Franciscan's residency program for training family practice doctors. In 2003, Grandview opened a $5.5 million osteopathic teaching facility and community health center to house its specialty teaching clinics and family practice program.

Under President Roy G. Chew, the hospital has won high marks in recent years. Solucient, the Evanston, Illinois-based health care information company, named Grandview one of the nation's "100 Top Hospitals" for 2001. And, in 2003, *U.S. News & World Report* listed it among America's Best Hospitals, ranking it thirty-fourth in the nation for the treatment of respiratory disorders and thirty-fifth for rheumatology.

As a result of the merger with Kettering, the Grandview Foundation was formed with a mission to provide grants and raise funds for osteopathic postgraduate education, health care, and research. In 2003, Kenneth B. Pugar, D.O., served as chairman.

MAHONING VALLEY GREEN CROSS HOSPITAL, WARREN

Occupying the former W. D. Packard mansion at 1320 Mahoning Avenue, N.W. in Warren, Mahoning Valley Green Cross Hospital opened on December 5, 1946. The nonprofit general hospital, with thirty-five beds and seven bassinets, was the project of four men: G. N. Mills, D.O., of Sharon, Pennsylvania; Harry E. Elston, D.O., of Niles, Ohio; John J. Mahannah, D.O., of Warren; and John S. Heckert, D.O., of Youngstown. William S. Konold served as consulting administrator; Florence Cody was resident manager.

In 1961, the hospital was forced to close its doors due to a mortgage foreclosure action. The following year, the nonprofit Community Hospital of Warren was founded. Incorporators were Anthony J. Candella, D.O., J. N. Cavalier, D.O., Robert P. Southard, D.O., William Borkowsky, D.O., Donald J. DelBene, Esq., and James G. Evans, hospital administrator. The new osteopathic hospital opened in the remodeled Packard mansion on June 12, 1962. A twenty-four-bed addition was completed in 1966, bringing capacity to sixty beds. In 1978, Warren Community Hospital announced that it would cease to operate as a hospital and instead expand into a one-hundred-bed nursing home. It continues to operate today.

NORTHEASTERN OHIO GENERAL HOSPITAL, MADISON

More than two thousand people turned out for the dedication of the new Northeastern Ohio General Hospital in Madison (Lake County) on August 30, 1959. It was the climax, said the *Buckeye*, of "many years of hard work and sometimes heartbreaking setbacks." Jessie M. Hutchison, D.O., of Geneva, conceived the project in the early 1950s, when the closest hospital open to a D.O. was located in Cleveland. A $20,000 bequest from a grateful patient marked the beginning of the campaign to build an osteopathic hospital to serve the rapidly growing Lake County. In 1954, Hutchison, together with Joseph S. Lefler, D.O., of Painesville (who would become chief of staff), and Charles Stull, D.O., of Geneva, formed a nonprofit corporation to build a twenty-five-bed hospital. The hospital was constructed at a cost of $507,112, assisted by a Hill-Burton grant and a bank loan. A year after opening, the hospital had served 1,410 patients.

A new $2.7 million wing, substantially financed by federal funds, was completed in 1976, raising capacity to eighty-two beds. But low occupancy and the resulting operating losses caused the hospital to close on December 1,

1988. Two weeks later, the new Lake Medical Center–Madison, a unit of Lake Hospital Systems, opened in the former Northeastern Ohio hospital building. The building later housed the east campus of Lakeland Community College.

ORRVILLE COMMUNITY OSTEOPATHIC HOSPITAL, ORRVILLE

With sixteen beds and eight bassinets, the nonprofit Orrville Community Osteopathic Hospital was formally dedicated on Sunday, December 13, 1953, when an open house attracted nearly two thousand persons. "Much sacrifice has gone into the making of the hospital," the *Buckeye* reported in January 1954, noting that the combined efforts of the community and the medical staff had resulted in $100,000 in gifts. Mervin Hostetler, president of the board of trustees, guided the "highly modernistic" one-story brick building to completion. The hospital's professional staff included R. J. Swoger, D.O., Alexander Zacour, D.O., and Edward Brown, D.O.; M. C. Kropf, D.O., served as medical director. In 1958, work began on a forty-nine-bed addition.

Following implementation of a new system of Medicare reimbursement in the 1980s, revenue declined. The struggling hospital was renamed Wayne General and Podiatry Hospital to better reflect the changing nature of its patient care. In 1986, the hospital consolidated with Dunlap Memorial Hospital. Today, the former osteopathic hospital houses a skilled-nursing facility.

PARKVIEW HOSPITAL, TOLEDO

In June 1943, the *Buckeye* announced that the osteopathic physicians of Toledo had purchased, for $50,000, the property of the late Judge John H. Doyle in the city's historic Old West End as the site for Toledo's first osteopathic hospital. The first officers were H. J. Long, D.O., president; Ralph D. Ladd, D.O., vice president; Leslie L. Billings, D.O., secretary; and V. W. Brinkerhoff, D.O., treasurer. William S. Konold served as consultant manager of the new hospital, which opened in June 1946 as a twenty-five-bed general hospital. Two years later, Wanda Gumm, resident manager, announced that the hospital had admitted 2,381 patients, performed 581 major operations and 632 minor ones, and assisted with the arrival of 304 newborns. Additional construction projects in 1954, 1957, 1962, and 1970 gradually increased bed capacity to 130.

Gilbert S. Bucholz, D.O., director of radiology and regional dean of the Ohio University College of Osteopathic Medicine, rewards the college's Dean Frank W. Myers, D.O., for helping to "raise the roof" for the expansion of Parkview Hospital, 1980. *Courtesy of Ohio Osteopathic Association.*

In 1980, under the leadership of Parkview CEO Bernhardt A. Zeiher, the hospital embarked on a $12.2 million expansion and renovation. By 1990, the facility had emerged as one of Toledo's principal acute-care medical centers, with an emphasis on personal, progressive care. Its staff of sixty osteopathic physicians and surgeons included Gilbert S. Bucholz, D.O., a past president of the American Osteopathic Association, and Michael Adelman, D.O., a past president of the American Osteopathic College of Proctology. But in 1994, as it competed with other area hospitals for diminishing inpatient business, Parkview filed a petition for bankruptcy in federal court and closed its doors. The hospital's teaching programs were acquired by nearby St. Vincent Mercy Medical Center, which also absorbed many Parkview doctors.

RICHMOND HEIGHTS GENERAL HOSPITAL,

RICHMOND HEIGHTS

Richmond Heights General Hospital traces its origin to Forest Hill Hospital, opened at 13240 Euclid Avenue in 1950 by ten Cleveland osteopathic physicians: Jerry A. Zinni, Domenic J. Aveni, Donald V. Hampton, Stanley B. Koerner, Leonard C. Nagel, Ann Prosen, H. W. Weichel, W. R. Disinger, Theodore F. Classen, and Leonard Greenbaum. Three years later, they purchased and substantially remodeled an emergency clinic located at 924 East 152nd Street in the heavily industrialized Collinwood neighborhood of Cleveland. The forty-three-bed hospital, representing an investment of $328,000, was financed in an unusual way: to create the

Jerry A. Zinni, D.O., is overcome with emotion during the dedication of Richmond Heights General Hospital, September 1961. Zinni led a group of Cleveland-area physicians—many of them colleagues on the staff of the old Forest Hill Hospital—to found and build the seventy-bed facility serving the city's eastern suburbs. Seated are James O. Watson, D.O. (with pipe), John R. Brostman, D.O., chief of staff, and local dignitaries. *Courtesy of Ohio Osteopathic Association.*

building fund, participating physicians contributed a percentage of their fees.

In 1959, physicians on the staff of Forest Hill, led by Dr. Jerry Zinni, purchased thirteen acres of land in suburban Richmond Heights and, following a protracted but successful zoning battle, moved ahead with plans to build a new osteopathic hospital. On September 25, 1961, patients in Forest Hill Hospital were moved to the new seventy-bed Richmond Heights General Hospital at 27100 Chardon Road. There, Zinni realized his dream of establishing a full-service osteopathic hospital with a respected medical education program and an emphasis on personalized, high-quality medical care. In 1969, a $2.5 million three-story addition, constructed in front of the existing hospital, added 120 beds, including a new 10-bed pediatric unit and 6-bed coronary care unit. A three-year, $20 million construction program completed in 1984 included a new and expanded emergency department and new surgical facilities.

In 1995, Richmond Heights General Hospital was acquired by Primary Health Systems and renamed Mt. Sinai Medical Center–East. Five years later, PHS filed for bankruptcy. In 2000, the hospital was purchased by University Hospitals Health System of Cleveland and reopened as UHHS Richmond Heights. It continues to operate as a teaching hospital affiliated with OU-COM. The Northeast Ohio Healthcare Foundation, formed as a result of the 1995 sale, has $5 million in assets. Chaired by Greg Nolfi, the foundation focuses on projects in Cuyahoga and surrounding counties. It has funded a mobile clinic staffed by residents to provide care in underserved areas and will provide $100,000 in scholarships to OU-COM students in the name of hospital founder Dr. Jerry Zinni.

SANDUSKY MEMORIAL HOSPITAL, SANDUSKY

In 1934, three Sandusky D.O.'s frustrated by the denial of admitting privileges at local hospitals—Lester R. Mylander, C. W. Koehler, and O. C. Ricelli—established a small hospital, initially sharing a cobblestone residence at 317 Fulton Street with the Loretta Britton Convalescent Home. Other area D.O.'s began to use the facility. When the convalescent home moved to new quarters, the expanded osteopathic facility was incorporated as the Hinde Memorial Hospital, taking its name from the Hinde Estate from which the building was leased. In 1940, Sandusky Memorial Hospital was incorporated as a nonprofit institution and granted a 5 percent allotment of the annual Erie County Community Fund drive to carry on its charitable work.

An open house on October 31, 1953, marked the completion of a new fifty-bed hospital at 2020 Hayes Avenue. A $500,000 addition in 1959 added thirty-two beds, and a new wing opened in 1964 brought Sandusky Memorial's bed count to 120. In 1976, the hospital broke ground for a $4.5 million expansion and modernization program that included twenty-five new beds, a coronary care unit, a dietary department, and additional space for laboratory, outpatient, and emergency departments. In 1985, Sandusky Memorial Hospital and Good Samaritan Hospital, an allopathic institution, consolidated to create the new joint-staff Firelands Community Hospital, an AOA- and JCAHO-accredited, 301-bed acute care facility. In 2001, Firelands purchased Providence Hospital in Sandusky. That facility now serves as the south campus of the renamed Firelands Regional Medical Center.

SELBY GENERAL HOSPITAL, MARIETTA

Selby General Hospital traces its roots to the Marietta Osteopathic Clinic, opened in 1927 as the project of four D.O.'s: Hubert L. Benedict, J. E. Wiemers, L. M. Bell, and J. D. Sheets. The same doctors, joined by A. Y. Siewers, organized a nonprofit hospital corporation in 1934. The clinic, which occupied a remodeled house at 304 Putnam Street, was expanded

Selby General Hospital moved to a new building on the outskirts of Marietta in 1965, when this photograph was taken. It has since expanded to serve a seven-county area of Ohio and West Virginia. In 2003, Selby was the only freestanding osteopathic hospital left in Ohio. *Photo by Leyfoto, courtesy of Ohio Osteopathic Association.*

with the construction of a two-story hospital at the rear. Marietta Osteopathic Hospital opened in July 1935. It was substantially enlarged in 1940 by an addition financed by oilman Francis M. Selby. In 1957, Mrs. William G. Selby of Sarasota, Florida, contributed funds allowing the hospital to purchase the clinic from the building's owners. The transaction allowed the hospital to move its administrative offices into the clinic and expand bed capacity. The hospital was renamed Selby General Hospital as a tribute to the late William G. Selby, the son of Francis Selby, and in recognition of the hospital's principal benefactors.

On April 25, 1965, a new Selby General Hospital with forty-four beds and four bassinets was dedicated on the outskirts of Marietta at the confluence of the Ohio and Muskingum Rivers. On January 14, 1973, two thousand guests attended as the hospital dedicated a new thirty-five-bed addition and ancillary facilities. In 1983, Selby broke ground for a $6.3 million expansion project that would affect "virtually every department in the hospital," according to hospital CEO Kenneth Malchiodi. Completed in 1985, the project included new intensive care and coronary care units, and expanded emergency, outpatient, obstetrics, and ultrasound departments. In 2002, Selby, which serves a five-county area of Ohio as well as West Virginia's Pleasants and Wood counties, was the only freestanding osteopathic hospital remaining in Ohio.

WARREN GENERAL HOSPITAL, WARREN

Warren General Hospital was dedicated on October 16, 1958, with a large crowd on hand to inspect the forty-two-bed osteopathic community facil-

ity. Located on Eastland Avenue in a quiet residential area, the $625,000 hospital was the project of three D.O.'s: William J. Timmins Jr., Eugene R. DeLucia, and Mario D. Massullo. In 1971, Ohio Gov. John J. Gilligan participated in ceremonies dedicating a $3.5 million four-story addition, which increased bed capacity to 174. In 1987, a cornerstone-laying ceremony marked the completion of a $10.4 million construction and renovation project that brought bed capacity to 212, added a new office building and registration lobby, and renovated much of the hospital's first floor.

St. Joseph Health Center, part of HM Health Services, acquired the hospital in 1996.

YOUNGSTOWN OSTEOPATHIC HOSPITAL, YOUNGSTOWN

In 1947, Arthur M. Friedman, D.O., and attorney Raymond Fine incorporated a new nonprofit hospital to serve the osteopathic physicians and surgeons in the Youngstown area. Not until five years later, however, after attorney Samuel H. Cooperman and Sol Leibel, D.O., joined the project, was the building program begun. Named after developer William Cafaro, a hospital benefactor, Cafaro Memorial Hospital (later renamed Youngstown Osteopathic Hospital) opened in March 1953 with thirty beds and six bassinets. J. Arnold "Jerry" Finer, D.O., headed a staff of twenty-one physicians and surgeons. By 1966, four major expansions—including the addition of a second floor—had increased capacity to 140 beds, with enlarged surgery, laboratory, and X-ray departments.

Cafaro Memorial Hospital (later renamed Youngstown Osteopathic Hospital) opened in 1953. Following several expansions that greatly increased capacity, Youngstown won designation by OU-COM as a regional teaching center in 1978, but was forced to close its doors in 2000. *Courtesy of Ohio Osteopathic Association.*

In 1978, following its designation by OU-COM as a regional teaching center for east central Ohio, Youngstown Osteopathic Hospital won a $2 million supplemental appropriation from the state for construction of a clinical teaching facility. In 1984, under the direction of CEO Mark C. Barabas, the hospital opened a $3.5 million addition housing new coronary, intermediate, maternity, and general medical units—changes that improved the quality of patient care and enhanced its role as a teaching hospital. Other innovations included a team approach to nursing care and a sleep lab for the diagnosis of sleep disorders.

For many years, Youngstown Osteopathic Hospital employees and physicians staffed a volunteer clinic at Youngstown State University, providing free health services to students. Alice T. Hill, the hospital's director of medical records and longtime executive secretary of the Youngstown Academy of Osteopathic Medicine, coordinated clinic operations. The hospital later won a contract with the university to staff an on-site student health service using staff physicians and family practice residents.

Declining admissions and revenue forced the hospital to close its doors in 2000.

INTRODUCTION

1. Information on the life of Andrew Taylor Still and the early development of osteopathy comes principally from Carol Trowbridge, *Andrew Taylor Still, 1828–1917* (Kirksville, Mo.: Thomas Jefferson University Press, 1991); and Norman Gevitz, *The D.O.'s: Osteopathic Medicine in America* (Baltimore: Johns Hopkins University Press, 1982).

2. John S. Haller Jr., *American Medicine in Transition, 1840–1910* (Urbana: University of Illinois Press, 1981), 98–99.

3. A. T. Still, *Autobiography of Andrew T. Still*, rev. ed. (Kirksville, Mo.: by the author, 1908), 258, 262.

4. Revised charter of October 30, 1894, quoted in Trowbridge, *Andrew Taylor Still*, 141.

5. The scientific study of medicinal drugs and their sources, preparation, and use.

CHAPTER I

1. Dr. Walter H. Siehl, "Osteopathy Comes to Ohio," typescript, files of the Ohio Osteopathic Association, Columbus, Ohio, n.p. The *Akron Official City Directory, 1897* (Akron: Burch Pub. Co., 1897), 301, lists "Eastman, Eugene H, osteopath, 113 S Broadway, bds [boards] Windsor Hotel."

2. Norman Gevitz, *The D.O.'s: Osteopathic Medicine in America* (Baltimore: Johns Hopkins University Press, 1982), 44–45.

3. "Honors Forty Year Men," *Buckeye Osteopathic Physician*, March 1940, 11; "Represent 102 Years of Practice," *Buckeye Osteopathic Physician*, August 1949, 9. The *Buckeye Osteopathic Physician* hereafter is referred to in the notes as *BOP*. Since most articles are short news items, reports, and the like, references in most cases are limited to the date of publication and page number.

4. "Honors Forty Year Men," *BOP*, March 1940, 11; "Dr. Katherine M. Scott Dies in California," *BOP*, May 1959, 5.

5. Quoted in Gevitz, *D.O.'s*, 36.

6. Jon F. Wills, "Early Ohio D.O.'s," *BOP*, October 1976, 9–10; "Dr. H. H. Gravett Dies," *BOP*, December 1958, 7.

7. "Man-Wife Team Retires after 35-Year Service in Ohio," *BOP*, October 1950, 7.

8. Legislation establishing district licensing boards beginning in 1811 was apparently ineffective and was repealed in 1833. William G. Rothstein, *American Physicians in the Nineteenth Century: From Sects to Science* (Baltimore: Johns Hopkins University Press, 1985), 339.

9. An act to regulate the practice of medicine in the state of Ohio, 92 Laws of Ohio 44 (1896).

10. Hart F. Page, *One Hundred Fifty Years of Service to Medicine, 1846–1996: The Ohio State Medical Association* (Columbus, Ohio: Ohio State Medical Association, n.d.), 71. For a brief account of the rise and fall of homeopathic and eclectic medicine, see James G. Burrow, "The Tale of Two Sects," in *Organized Medicine in the Progressive Era: The Move toward Monopoly* (Baltimore: Johns Hopkins University Press, 1977), 71–87.

11. This account of the early legal battles in Ohio is largely based on E. R. Booth, *History of Osteopathy, and Twentieth-Century Medical Practice* (Cincinnati: Jennings and Graham, 1905), 131–38, 173–76.

12. 92 Laws of Ohio 44, 47 (1896).

13. A copy of the letter can be found in a scrapbook kept by Hugh H. Gravett, D.O., now the property of the Ohio Osteopathic Association (hereafter, OOA), Columbus, Ohio.

14. A method of objecting that admits the facts of the opponent's argument but denies that they sustain the pleading based upon them.

15. Gravett scrapbook, OOA.

16. An act to amend . . . an act entitled "An act to regulate the practice of medicine in the state of Ohio," 94 Ohio Laws 197, 201 (1900).

17. M. F. Hulett, "Veteran Fighter for Osteopathy Tells of the Earlier Days," *BOP*, May 1937, 14.

18. This account of the Gravett case is based on news clippings contained in the doctor's own scrapbook, cited in note 13 above.

19. In 1947, Dr. Hugh H. Gravett was named an honorary life member of the American Osteopathic Association. Three years later, the Ohio Osteopathic Association similarly honored him.

20. Articles of Incorporation of the Ohio Osteopathic Society, December 7, 1901, files of the OOA.

21. Hildreth's report on the 1902 legislative battle is contained in Siehl, "Osteopathy Comes to Ohio," n.p.

22. An act to amend section 4403ƒ of the Revised Statutes of Ohio, 95 Ohio Laws 212 (1902).

23. Hildreth report, in Siehl, "Osteopathy Comes to Ohio," n.p.

CHAPTER 2

1. Dr. Walter H. Siehl, "Osteopathy Comes to Ohio" typescript, files of the OOA, Columbus, Ohio, n.p. Siehl, who served as president of the OOA in 1948–49, took an active interest in collecting and preserving the early history of osteopathy in Ohio, writing two manuscripts, "A Brief History of Osteopathy in Ohio" and "Osteopathy Comes to Ohio." The manuscripts were never published.

2. E. R. Booth, *History of Osteopathy, and Twentieth-Century Medical Practice* (Cincinnati: Jennings and Graham, 1905), 97–98.

3. Siehl, "Osteopathy Comes to Ohio," n.p.

4. A copy of the directory can be found in the files of the OOA. By comparison, the Ohio State Medical Association counted 3,866 members in 1911, with 87 component (county) societies. James G. Burrow, *AMA: Voice of American Medicine* (Baltimore: Johns Hopkins Press, 1963), 401.

5. The national organization of osteopathic physicians, founded in 1897, was known as the American Association for the Advancement of Osteopathy until 1901.

6. C. M. Turner Hulett, D.O., "The Grace of Persistence," *Journal of the American Osteopathic Association* 9 (May 1910): 389–91; Norman Gevitz, *The D.O.'s: Osteopathic Medicine in America* (Baltimore: Johns Hopkins University Press, 1982), 54.

7. "Osteopathy in War," *Journal of the American Osteopathic Association* 16 (July 1917): 1278.

8. "Program for All Ohio Day, Monday, August 6, 1917, part of the regular program of the AOA National Convention at Columbus, Ohio, August 6–10, 1917," contained in a scrapbook kept by Hugh H. Gravett, D.O., now the property of the OOA, Columbus, Ohio.

9. *BOP,* January 1917, n.p.

10. Siehl, "Osteopathy Comes to Ohio," n.p.

11. *BOP,* September 1933, 8. Morris Fishbein, M.D., was the longtime editor of the *Journal of the American Medical Association.* In *The Medical Follies* (New York: Boni & Liveright, 1925), 58–59, Fishbein wrote that osteopathy "is essentially an attempt to get into the practice of medicine by the back door."

12. *BOP,* November 1933, 3.

13. Ibid., 2.

14. *BOP*, October 1933, 3.

15. *BOP*, January 1934, 1.

16. In 1873, the first American hospital survey located 178 hospitals. On January 1, 1923, there were 4,978 hospitals in the United States. Charles E. Rosenberg, *The Care of Strangers: The Rise of America's Hospital System* (New York: Basic Books, 1987), 341.

17. E. R. Booth, *History of Osteopathy, and Twentieth-Century Medical Practice*, memorial ed. (Cincinnati: Caxton Press, 1924), 745.

18. "Fine New Delaware Springs Sanitarium Is Auspiciously Opened with Public Reception," *Delaware Semi-Weekly Gazette*, March 17, 1916, 6.

19. *BOP*, December 1936, 5. The Jane M. Case Hospital acquired the Delaware Springs Sanitarium in 1928. In the early 1930s, it briefly leased one wing of the facility as a private osteopathic sanitarium. In 1947, in an unusual cooperative agreement, the Delaware Osteopathic Hospital was established on the first floor of the Case Hospital. It operated until 1962. In 1972, Jane M. Case Hospital became known as Grady Memorial Hospital. See Ray Buckingham, "75 Years of Service: Grady Memorial Hospital, 1904–1979," special supplement to *Delaware Gazette and Sunbury News*, August 8, 1979.

20. Booth, *History of Osteopathy*, memorial ed., 746–47.

21. *BOP*, December 1934, 4; March 1940, 8–9.

22. *Buckeye Osteopath*, December 1934, 4; *BOP*, February 1940, 1, 6–7; January 1958, 3.

23. *BOP*, November 1938, 1.

24. James G. Monnett Jr., "Old Hall Home Bought for Clinic," *Cleveland Plain Dealer*, reprinted in *Buckeye Osteopath*, April 1935, 6; *BOP*, March 1942, 8.

25. "The Hayes-Mayberry Osteopathic Hospital, East Liverpool, Ohio," files of the OOA.

26. *BOP*, May 1940, 1; July 1940, 9; Aug. 1940, 5.

27. *BOP*, May 1937, 12.

28. *BOP*, May 1935, 7.

29. John W. Keckler, D.O., to L. D. Leidheiser, D.O., April 20, 1973.

30. OOHA Constitution and Bylaws, files of the OOA.

31. *BOP*, September 1939, 6; November 1936, 7.

32. "Free Clinic Announcement by Dr. Sheppard," *Buckeye Osteopath*, October 1933, 6.

33. *Buckeye Osteopath*, April 1934, 1.

34. Letter to the editor, *Buckeye Osteopath*, May 1935, 8.

35. "Child Health Clinic Conducted at Fair," *BOP*, September 1939, 1, 7.

36. Dr. C. A. Purdum, "Can You Give a Chiropractic Treatment?" *BOP*, November 1939, 2.

37. "Ohio Society Starts Work on Osteopathic Participation in Social Security Measures," *BOP*, March 1938, 4–5. The Burke-Drew Bill, eliminating discrimination against osteopathic physicians by the U.S. Employment Compensation Committee, was signed into law by President Franklin D. Roosevelt on May 31, 1938.

38. Obituary of Dr. Raymond P. Keesecker, *BOP*, June 1960, 12. Keesecker left Cleveland in 1951 to become editor of AOA publications.

39. *BOP*, November 1936, 1.

40. Board of Trustees Minutes, May 16, 1933, files of the OOA.

41. Board of Trustees Minutes, September 10, 1933, files of the OOA.

42. In December 1936, page one of the *Buckeye Osteopathic Physician* reported the number of osteopathic physicians then practicing in each state district: Toledo, 57; Cleveland, 99; Akron, 67; Columbus, 77; Dayton, 87; Cincinnati, 33; and Marietta, 21.

43. Board of Trustees Minutes, May 7, 1929, files of the OOA.

44. Board of Trustees Minutes, May 13, 1934, files of the OOA.

45. Report of Secretary M. A. Prudden, D.O., for 1934–35, Board of Trustees Minutes, files of the OOA.

46. *BOP*, September 1937, 1; James O. Watson, "Osteopathy in Ohio as a Big Business," *BOP*, December 1937, 1, 6.

47. Board of Trustees Minutes, January 9, 1938, files of the OOA. According to the minutes, "Each trustee said their district was against it at this time."

48. *BOP*, June 1939, 2, 6.

49. The Treasurer's Report of the Ohio Society of Osteopathic Physicians and Surgeons, as of December 6, 1939, showed income of $3,822 and expenditures of $3,681, leaving a balance of $141.

50. *BOP*, June 1938, 7.

51. *BOP*, October 1939, 1.

CHAPTER 3

1. Board of Trustees Minutes, December 12, 1939, files of the OOA, Columbus, Ohio. According to the minutes, on the matter of hiring an executive secretary, President Licklider "told the board of the Medical [Society] set up in Ohio, with its full time executive secretary and two public relations men under him. He stated that he felt the board had a decision to make that would either advance Osteopathy in Ohio, or face the possibility of a bad set back."

For an appreciation of Licklider's career, see "Dr. Licklider Dies, Doctors Hospital Co-Founder," *BOP*, June 1988, 11.

2. *BOP*, November 1940, 11.

3. "Twenty-nine Years of Loyal Opposition," address to the Society of Divisional Secretaries, annual meeting of the AOA House of Delegates, Chicago, Ill., July 21–23, 1968, property of Mary Jane Carroll, Columbus, Ohio.

4. *BOP*, March 1940, 2.

5. Board of Trustees Minutes, March 3, 1940, files of the OOA; *BOP*, March 1940, 1.

6. In June 1973, by another vote of its members, the Ohio Osteopathic Association of Physicians and Surgeons changed its name to Ohio Osteopathic Association.

7. Board of Trustees Minutes, January 12, 1941, files of the OOA.

8. Certificate of Amendment, files of the OOA; *BOP*, December 1940, 2; Board of Trustees Minutes, January 12, 1941, files of the OOA; *BOP*, May 1941, 2.

9. Board of Trustees Minutes, January 10, 1937, files of the OOA. Watson was succeeded as chairman by Edgar Q. Lamb, D.O., in 1938 but remained an active member of the committee.

10. Board of Trustees Minutes, May 17, 1938, files of the OOA. At the meeting, the board voted to hire Hamer for one year at a salary of $1,500.

11. Norman Gevitz, *The D.O.'s: Osteopathic Medicine in America* (Baltimore: Johns Hopkins University Press, 1982), 81. As Gevitz explains, in some of these states D.O.'s continued to be ineligible if they were unable to meet mandated requirements for preprofessional college work or internships.

12. *BOP*, February 1935, 1.

13. In 1956, O'Neill was elected governor of Ohio. Two years later, he was appointed chief justice of the Ohio Supreme Court.

14. The amendment to Senate Bill No. 181 is reported in *BOP*, April 1939, 2.

15. J. O. Watson, *Further Recollections with Respect to the Development of Doctors Hospital and Other Commentary* (Columbus, Ohio: privately printed, 1984), n.p.

16. By 1972, eleven D.O.'s were serving as county coroners in Ohio.

17. Clarence L. Corkwell to Ohio Osteopathic Physicians and Surgeons, May 4, 1940.

18. Board of Trustees Minutes, October 1, 1939, files of the OOA.

19. "All Out for Civilian Defense," *BOP*, January 1942, 1.

20. According to Hart F. Page, *One Hundred Fifty Years of Service to Medicine, 1846–1996: The Ohio State Medical Association* (Columbus: Ohio State Medical Association, n.d.), 26, some 3,120 members of the Ohio State Medi-

cal Association served as medical officers in the armed forces during World War II.

21. *BOP*, February 1941, 3.

22. *BOP*, May 1941, 3.

23. J. O. Watson, D.O., to Jon Wills, April 3, 1980.

24. Ibid.

25. Since 1943, six osteopathic physicians have served on the State Medical Board of Ohio: James O. Watson, D.O., of Columbus, 1943–1972; William J. Timmins Jr., D.O., of Warren, 1972–1975; Evelyn L. Cover, D.O., of Columbus, 1975–1983; John E. Rauch, D.O., of Logan, 1983–1990; Theresa M. Hom, D.O., of Columbus, 1990–1993; and Anita M. Steinbergh, D.O., of Westerville, 1993 to the present. Cover was the first woman to be appointed to the state medical board and the first D.O. to serve as president. Rauch and Steinbergh also served terms as president. The State Medical Board of Ohio licenses and registers medical doctors, doctors of osteopathic medicine, doctors of podiatric medicine, physician assistants, anesthesiologist assistants, massage therapists, cosmetic therapists, acupuncturists, mechanotherapists, and naprapaths. It also enforces disciplinary laws relating to licensees as enacted by the Ohio General Assembly.

26. Beginning in 1968, osteopathic applicants took the same examination as all other graduates.

27. J. O. Watson, D.O., to Jon Wills, April 3, 1980.

CHAPTER 4

1. *BOP*, May 1945, 1.

2. Richard L. Sims, "The Centennial History of the Ohio Osteopathic Association, 1898–1999," typescript, files of the OOA, Columbus, Ohio, 18. The first OOA House of Delegates met on May 11, 1946.

3. *BOP*, July 1947, 4.

4. *BOP*, July 1949, 6.

5. *BOP*, December 1949, 12.

6. *BOP*, June 1954, 3.

7. Sims, "Centennial History," 29. In 1959, Dr. Charles L. Naylor, chairman of the OPF Committee of the OOA, reported that the 1958–59 fund drive raised a record $55,356 from Ohio D.O.'s for the support of the colleges. *BOP*, March 1959, 6.

8. *BOP*, November 1959, 2.

9. *BOP*, February 1953, 13.

10. *BOP*, September 1957, 18.

11. *BOP,* October 1956, 4, 6; August 1958, 2.

12. See Norman Gevitz, "The California Merger," in *The D.O.'s: Osteopathic Medicine in America* (Baltimore: Johns Hopkins University Press, 1982), 99–116. The College of Osteopathic Physicians and Surgeons became the California College of Medicine. It became part of the University of California system in 1962 and relocated to a new campus at Irvine several years later.

13. *BOP,* May 1961, 4–7.

14. "Trustees Statement on A.M.A. Policy," *AOA News Bulletin* 4 (July 1961): 3, cited in Gevitz, *D.O.'s,* 119.

15. William S. Konold to Ohio Osteopathic Association Members, January 19, 1962.

16. "Osteopaths Snub AMA Peace Move," *Cleveland Plain Dealer,* May 23, 1967.

17. "We Stand United," *BOP,* August 1967, n.p.

18. *BOP,* May 1971, 38. Among those called was Dr. James Sosnowski, a resident at Doctors Hospital, Columbus. He was killed in action in Vietnam on February 16, 1968, becoming the first D.O. to give his life for his country.

19. Sims, "Centennial History," 46.

20. Hart F. Page, *One Hundred Fifty Years of Service to Medicine, 1846–1996: The Ohio State Medical Association* (Columbus: Ohio State Medical Association, n.d.), 78.

21. In 1996, six hundred Ohio D.O.'s were dues-paying members of the OSMA, according to Page, *One Hundred Fifty Years of Service to Medicine,* 78.

22. In 1943, over two hundred doctors attended a two-and-a-half-day "refresher" course. *BOP,* November 1943, 2.

23. Reprinted in *BOP,* June 1949, 12.

24. "President's Message," *BOP,* January 1970, 3.

25. "President's Message," *BOP,* May 1970, 2.

26. "AOA President's Message," *BOP,* July 1979, 19.

27. The Medicare and Medicaid programs were formally enacted in 1965 as amendments (Titles XVIII and XIX, respectively) to the Social Security Act (1935).

28. *BOP,* July 1968, 4.

29. Gevitz, *D.O.'s,* 95.

30. "President's Message," *BOP,* November 1960, 4.

31. *BOP,* December 1968, 1–2, 3. For a full account of the image study, see James F. Engel, W. Wayne Talarzyk, and Carl M. Larson, *Cases in Promotional Strategy* (Homewood, Ill.: Richard D. Irwin, 1971).

32. *BOP,* August 1970, n.p.

CHAPTER 5

1. *BOP*, July 1973, 12.

2. "Ohio in the Colleges," *BOP*, February 1938, 1, 6.

3. *BOP*, February 1960, 6.

4. In 1963, OOA president Donald Siehl, D.O., reported that the ratio of D.O.'s to the U.S. population had declined, from 9.4 per 100,000 in 1940 to 7.8. *BOP*, February 1963, 3.

5. *BOP*, October 1969, 1.

6. The first class included an Ohioan, James S. Lapcevic.

7. *BOP*, January 1974, n.p.

8. *BOP*, June 1975, n.p.

9. A copy of the resolution appears in *BOP*, July 1975, n.p.

10. "Osteopaths Back Malpractice Bill," *Cleveland Press*, June 10, 1975.

11. "No to Med School," June 9, 1975.

12. *Athens (Ohio) Messenger*, June 22, 1975, reprinted in the *BOP*, July 1975, n.p.

13. "Inaugural Address by Martin E. Levitt, D.O.," *BOP*, July 1975, n.p.

14. J. O. Watson, D.O., to Jon Wills, October 5, 1981.

15. Carl J. Denbow, "The Tenth Is Ten," *BOP*, August 1986, 8.

16. Harry B. Crewson to Dr. Charles J. Ping, Board of Trustees, [and] Senior Administrators, August 26, 1975.

17. Commencement Address, Ohio University College of Osteopathic Medicine, June 2, 2001. A copy is in the files of the OOA, Columbus, Ohio.

18. *BOP*, January 1976, 14–15.

19. *BOP*, January 1983, 24.

20. In 1999, Harold C. Thompson III, D.O., assistant professor of emergency medicine in the Department of Family Medicine, succeeded Myers as host. Today, *Family Health* is heard daily in 175 nations around the world, as well as on the Armed Forces Radio Network.

21. Sponsored by the American Association of Colleges of Osteopathic Medicine and a consortium of twelve osteopathic medical schools, Summer Scholars was continued by OU-COM even after the federally funded program was moved to the Texas College of Osteopathic Medicine after 1984.

22. *BOP*, April 1991, 16.

23. *BOP*, December 1993, 23.

24. *BOP*, May 1993, 18.

25. By 2001, there were four: Northwest CORE, comprising Firelands Regional Medical Center in Sandusky, St. John West Shore Hospital in Westlake,

and St. Vincent Mercy Medical Center in Toledo; Northeast CORE, comprising Cuyahoga Falls General Hospital, Doctors Hospital of Stark County, St. Joseph Health Center in Warren, South Pointe Hospital–Cleveland Clinic Health System in Warrensville Heights, and University Hospitals Health System–Richmond Heights Hospital; Southwest CORE, comprising Grandview Hospital in Dayton and Southern Ohio Medical Center in Portsmouth; and Southeast CORE, comprising Doctors Hospital of Columbus–OhioHealth and O'Bleness Memorial Hospital in Athens.

26. Brian O. Phillips and Christopher Duffrin, "Ohio Osteopathic Network of Excellence: Establishing a Statewide Telehealth Consortium," *Journal of the American Osteopathic Association* 101 (December 2001): 720–22.

27. Jon F. Wills (executive director, OOA), in discussion with the author, July 2002.

28. Ohio University College of Osteopathic Medicine, *COM FACTS*, fact sheet issued May 2001. A copy is in the files of the OOA.

29. Ohio University College of Osteopathic Medicine, *The Primary Choice: Ohio University College of Osteopathic Medicine, 2001–2002,* ([Athens, Ohio]: n.d.), 2.

30. Richard L. Sims, "Centennial History of the Ohio Osteopathic Association, 1898–1999," typescript, files of the OOA, 58.

CHAPTER 6

1. *BOP,* December 1979, 12.

2. *BOP,* August 1978, 7.

3. "Health Planner Rejects Osteopaths' Challenge," *Cleveland Plain Dealer,* August 8, 1970, 9A. In 1978, Bay View and St. John Hospitals contracted to jointly build a new health care complex in suburban Westlake. Bay View medical staff and personnel moved to the new St. John West Shore Hospital, opened in 1981.

4. *BOP,* March 1978, 15.

5. "Citizens Attack MHPC Hospital Plans," *Painesville (Ohio) Telegraph,* December 8, 1978, clipping in files of the OOA, Columbus, Ohio.

6. In 1995, the Ohio General Assembly repealed the state's certificate-of-need program, except for long-term care facilities.

7. See Hugh M. Culbertson and Guido H. Stempel III, "OOA Survey Report: The Public View of Osteopathic Medicine in Ohio," *BOP,* December 1982, 8–19. For an extended treatment, see Hugh M. Culbertson, Carl J. Denbow, and Guido H. Stempel III, "David and Goliath Coexist: The Story of

Osteopathic Public Relations," in Hugh M. Culbertson, Dennis W. Jeffers, Donna Besser Stone, and Martin Terrell, *Social, Political, and Economic Contexts in Public Relations: Theory and Cases* (Hillsdale, N.J.: Lawrence Erlbaum Associates, 1993), 227–69.

8. *BOP*, June 1984, 7.

9. *BOP*, June 1987, 16; July 1988, 15.

10. "NEO Hospital Still Seeks Merger, but Hopes Dim," *Willoughby (Ohio) News-Herald*, November 12, 1988, A5.

11. *BOP*, February 1990, 9; April 1990, 14.

12. *BOP*, February 1991, 8; August 1991, 11.

13. *BOP*, May 1992, 2.

14. *BOP*, September 1992, 2.

15. *BOP*, June 1989, 4.

16. *BOP*, April 1990, 10; October 1990, 4.

17. *BOP*, July 1992, 4.

18. *BOP*, Summer 1997, n.p.

19. Discussion with author, July 2002.

20. *BOP*, Summer 1997, n.p.

21. Peter Bell, D.O., e-mail message to Jon F. Wills, October 9, 2001.

22. Richard A. Vincent to Jon F. Wills, April 24, 2002.

23. "Presidential Inaugural Address," *BOP*, Summer 2000, 19.

24. Terri Kovach, executive director, Brentwood Foundation, e-mail message to author, September 24, 2003.

25. Jon F. Wills, e-mail message to author, September 24, 2003.

26. "Presidential Inaugural Address," *BOP*, Summer 1999, 14.

27. Norman Gevitz, *The D.O.'s: Osteopathic Medicine in America* (Baltimore: Johns Hopkins University Press, 1982), 141.

28. Gunnar B. J. Andersson, M.D., Ph.D., Tracy Lucente, M.P.H., Andrew M. Davis, M.D., M.P.H., Robert E. Kappler, D.O., James A. Lipton, D.O., and Sue Leurgans, Ph.D., "A Comparison of Osteopathic Spinal Manipulation with Standard Care for Patients with Low Back Pain," *New England Journal of Medicine* 341 (November 4, 1999): 1426–31.

29. Joan Wehrle, State Medical Board of Ohio, e-mail message to Jon F. Wills, July 23, 2003.

30. *BOP*, Summer 1999, 3.

A NOTE ON SOURCES AND SELECTED BIBLIOGRAPHY

This history relies principally on the records and publications of the Ohio Osteopathic Association and its predecessors. Especially valuable were the *Buckeye Osteopath*, published intermittently between 1923 and 1936, and its successor, *Buckeye Osteopathic Physician*, published continuously since 1936. Both publications constitute a detailed chronicle of events, trends, people, and institutions of the osteopathic medical profession in Ohio. Minutes of the meetings of the board of trustees were another important source; unfortunately, those for the period prior to 1928 have been lost, and the author has had to construct the story of the early years of osteopathy in Ohio from, quite literally, fragments. A scrapbook kept by Hugh H. Gravett, D.O., of Piqua, containing news clippings, letters, and other documents for the period 1897–1918, provided valuable insight into the early struggle for osteopathic recognition. Unpublished histories of osteopathic medicine in Ohio prepared by two key players in the story—Walter H. Siehl, D.O., of Cincinnati, and Richard L. Sims, former executive director of the OOA—provided general overviews of the profession as it developed in Ohio.

Also useful were the transcript of a 1977 interview conducted by Carl J. Denbow with James O. Watson, D.O., and several letters Watson wrote to OOA Executive Director Jon F. Wills reflecting on his experiences in the profession, especially the legislative battles. Two speeches by former OOA Executive Secretary William S. Konold—one given in 1968 to the Society of Divisional Secretaries, the other in 1971 to the Association of Osteopathic State Executive Directors—provided a valuable perspective on the profession's development and struggles. Personal interviews with Mary Jane Carroll (daughter of William S. Konold), John P. Sevastos, D.O., Richard L. Sims, and Jon F. Wills aided the author's understanding of trends and, especially, personalities important to the story of osteopathy in Ohio. Finally, the following published works also proved useful.

Booth, E. R. *History of Osteopathy, and Twentieth-Century Medical Practice.* Cincinnati: Jennings and Graham, 1905.

————. *History of Osteopathy, and Twentieth-Century Medical Practice.* Memorial Edition. Cincinnati: Caxton Press, 1924.

Brentwood Foundation. *A Touch of Magic at Brentwood: The History of an Osteopathic Hospital.* Cleveland: Brentwood Foundation, 2000.

Burrow, James G. *Organized Medicine in the Progressive Era: The Move toward Monopoly.* Baltimore: Johns Hopkins University Press, 1977.

Carroll, Mary Jane. *Doctors Hospital, 1940–1990: The First Fifty Years.* Columbus: Doctors Hospital, 1990.

Gevitz, Norman. *The D.O.'s: Osteopathic Medicine in America.* Baltimore: Johns Hopkins University Press, 1982.

Still, A. T. *Autobiography of Andrew T. Still.* Rev. ed. Kirksville, Mo.: by the author, 1908.

Trowbridge, Carol. *Andrew Taylor Still, 1828–1917.* Kirksville, Mo.: Thomas Jefferson University Press, 1991.

Watson, J. O. *Recollections of Dr. J. O. Watson on the History of Doctors Hospital.* Columbus: privately printed, 1976.

————. *Further Recollections with Respect to the Development of Doctors Hospital and Other Commentary.* Columbus: privately printed, 1984.

INDEX

Numbers in italic indicate pages on which pictures of individuals appear.

Candella, Anthony J., 133
Carroll, Mary Jane, 68–69
Cavalier, J. N., 133
Centers for Osteopathic Research and Education (CORE), 78, 83, 95, 99
certificate-of-need regulations, 87–89
Chastang, Charles J., 48, *49*, 70
Chew, Roy G., 106, 132
Chila, Anthony G., *79*
Chinese medicine, 107
Clarey, Alison A., 109, *109*
Classen, Theodore F., 106, 124, *124*, 125, 135
Cleveland Academy of Osteopathic Medicine, 60, 88, 99
Cleveland Clinic Health System, 98, 99, 104, 105, 106, 125, 128–29
Cleveland Osteopathic Clinic, 57
Cleveland Osteopathic Hospital, 27–28, 32, 57, 123
Cluett, Therese, 5, 6–7
Clybourne, Harold E., 29, *51*, *67*, 126
Clybourne, Mildred, 45
Coan, James E., 103
Coan, John J., 23
Cody, Florence, 133
College of Osteopathic Physicians and Surgeons, Los Angeles, 56, 57, 71
Columbus Radium Hospital, 35
Committee on Maternal and Child Health, 31–32
Community Hospital of Warren, 133
Connelly, Elizabeth, *96*
Cooperman, Samuel H., 139
CORE. *See* Centers for Osteopathic Research and Education (CORE)
Corkwell, Clarence, L., 19, 23, 33, 43
Costin, J. Richard, *108*
Cover, Evelyn L., *89*
Crewson, Harry B., 74
Culbertson, Hugh M., 89
Cuyahoga Falls General Hospital, 99, 125–26

Day, June B., 51
Dayton Osteopathic Hospital, 25, 131
Delaware Springs Sanitarium, 20, 24–25, 144n19
Del Bane, Michael, 72
DelBene, Donald J., 133
DeLucia, Eugene R., 65, 139
DeVoe, Gretchen, 87
Dilatush, Frank A., 19, 25, 131
Dill, Heber M., 25, 131
Dill-Dilatush Clinic, 25

Dilling, Thomas, *108*
Disinger, W. R., 135
district academies, 39, 51, 109
Dobeleit, Richard F., 25, 131
Doctor of Osteopathy (D.O.) degree, 3
Doctors Hospital (Columbus), 38
    conversion to outpatient and urgent-care facility, 98–99
    history of, 29, 126–28
    open-heart surgery approved for, 88
    postwar growth of, 53, 78, 79
    and Watson, James O., 35–36
Doctors Hospital of Nelsonville, 79, 127
Doctors Hospital of Stark County, 54, 98, 107, 128–29
Doctors [Hospital] West, 53, 126–27
Drevenstedt, Jean, *96*
Drew, Ira W., 101
Dunigan, George F., Jr., 70, 72, 92, *97*

East Liverpool Osteopathic Hospital, 29, 129–30
Eastman, Eugene H., 5, 8
Elliott, Chip, 68–69
Elston, Harry E., 20, 133
Engel, James F., 67–68
Evans, James G., 133
Eversull, Warner S., 20

Fahlgren and Ferriss, 89
Family Health, 81
Faverman, Gerald A., 74–76, *75*, 77, *77*, *78*
Fine, Raymond, 139
Finer, J. Arnold, Jr., 128, 139
Firelands Regional Medical Center, 137
Foraker, Joseph B., 9
Forest Hill Hospital, 104, 135–36
Foster, E. Lee, *108*, *109*, 110–11
Franciscan Medical Center, 132
free child health clinics, 31
free clinic committee, 30
Friedman, Arthur M., 139
Fries, Tom, 72, 73, *73*, *101*
Fulton, Richard L., 64

George, Robert J., 94
Gevitz, Norman, 83
Giddings, Helen Marshall, 5, 46
Gilligan, John J., 71, 139
Gilmore, Paul, 72
Glass, O. R., 34
Glenn, John, 82, *101*
Goldberg, David D., *108*